Pelican Books

THE PLANETS – A Decade of Discovery

Peter Francis was born in Northern Rhodesia (Zambia) in
1944 and educated in Britain. He graduated from Imperial
College, London, in 1966, obtained his Doctorate in 1969,
and is now a lecturer in Earth Sciences at the Open
University. He has combined a life-long interest in the
solar system with a hectic career in volcanology. After a
teenage involvement with lunar and planetary observations
with home-made telescopes, he concentrated on terrestrial
geology and has devoted the last decade to studies of the
volcanoes of the Andes. The present era of spacecraft
exploration of the planets and the flood of new planetary
data rekindled his old enthusiasm, and the present book is
the result.

Peter Francis is also the author of *Volcanoes* (Pelican, 1976).

Peter Francis

THE PLANETS
A DECADE OF DISCOVERY

Penguin Books

Penguin Books Ltd, Harmondsworth, Middlesex, England
Penguin Books, 625 Madison Avenue, New York, New York 10022, U.S.A.
Penguin Books Australia Ltd, Ringwood, Victoria, Australia
Penguin Books Canada Ltd, 2801 John Street, Markham,
Ontario, Canada L3R 1B4
Penguin Books (N.Z.) Ltd, 182–190 Wairau Road,
Auckland 10, New Zealand

First published 1981

Designed by Philip Hall
Diagrams drawn by Raymond Turvey

Filmset, printed and bound in Great Britain
by Hazell Watson & Viney Ltd, Aylesbury, Bucks
Set in Monophoto Bembo

For my parents

Contents

Author's Note   Except where indicated otherwise in the
captions, all lunar and planetary photographs reproduced
in the book are provided by courtesy of NASA and
the National Space Sciences Data Center.

Throughout the text, wherever figures are given in
billions, these accord with the now internationally accepted
US billion (1,000 million), not the former British billion
(1 million million).

1. Groundwork

We live in troubled times. On the one hand, the world is faced with the massive physical problems of keeping its swelling population supplied with food and energy, and of maintaining an acceptable environmental balance in the process. On the other, deep-seated social unrest affects a large proportion of the world's population as conflicting political and religious ideologies struggle for dominance. One of the factors contributing to this unhappy situation has been the headlong rate of technological advance during this century, which has led to some countries achieving high standards of material wealth, while others continue to struggle along in the same way as they have done for centuries: in poverty, disease and famine.

In these circumstances, it is easy to lose sight of the fact that we are living through one of the greatest periods of scientific advance in the history of civilization; one that is purely the result of technological progress. It is important here to distinguish between 'science' and 'technology'. Most of the things that we associate with advance in modern civilization – such as transistor radios and pocket calculators – are the results of technological achievements rather than advances in *knowledge*. The pocket calculator exploded on the market in 1970, not because of any sudden discovery about the principles of calculation, or even of electronics, but because technology made it possible to manufacture the essential microelectronic 'chips' in huge numbers at low cost. Science, however, is concerned only with knowledge – it *is* knowledge – and in itself is neither good nor bad, economic nor uneconomic.

In 1957, the first artificial satellite, Sputnik 1, was put into orbit around the Earth. By 1978, technology had advanced so far that men had stood on the Moon, spacecraft had landed on Mars and Venus, and others had visited Mercury and Jupiter. Thus, in two decades, our knowledge of the solar system advanced many times further than it had over the previous three millennia. In the space of a few years, answers were found to problems which had confronted Man since he first looked up to the heavens.

The saddest aspect of this glorious flowering of science is that it is passing almost unnoticed by the public at large. True, the manned Apollo landings were accompanied by almost hysterical publicity, and a large fraction of the world's population was glued to its television sets to watch the first landing. But it was the fact of making the landing that caught people's imagination; they respected the achievement in the same way as the climbing of Everest. Little or no attention was paid to the scientific discoveries, and by the end of the Apollo landings, a scant three years after the first, public interest had waned. In the space of only three years, live television from the Moon had become a commonplace. Public interest flared up again during the time of the Viking landings on Mars, but only briefly. Perhaps because so many major developments came hard on each other's heels, many major missions went entirely unnoticed. Everybody knows that the Americans successfully landed men on the Moon, but how many know that the Russians have been able to land unmanned spacecraft on the Moon, obtain samples, and return them to Earth?

The news media have a large measure of responsibility for this unbalanced situation, since, as a matter of policy, they concern themselves almost exclusively with 'human interest' stories. They are much more likely to print a report on what an astronaut has for breakfast, or how he disposes of his solid wastes, than one dealing with what the Moon is made of. The scientific community is also guilty to some degree, since, while scientists shower each other with a rain of turgid and repetitive technical papers, they ignore their duty to keep the public (who pay for their work through taxes) informed of their activities, and, with a few exceptions, make no attempt at all to share the excitement of their discoveries with the public.

It is also ironic that the present burgeoning of our knowledge of the solar system comes at a time when the average man has never had less direct experience of the stars and planets. A century ago, a shepherd in his field could probably recognize several of the principal constellations, and would certainly have been able to spot the major planets. Most modern city dwellers would find it impossible to identify *any* of the planets, and many probably live their entire lives without ever consciously seeing one. Not only do the fume-polluted air and the glare of street lights make it physically difficult for them to observe the night sky, but it is also unnecessary for them to do so. City dwellers have no need to use the Sun and stars to monitor the progress of the seasons, and even those who dabble in astrology would never dream of actually going out to observe for themselves the alignments of stars and planets which they claim rule their lives.

As a result of factors like these, we are living through an epic period in the history of human achievement, but one which only a small handful of scientists is appreciating to the full. Centuries from now, our descendants will be talking of this period as a major landmark in the advance of knowledge. The tragedy is that so few of us are yet able to recognize that landmark. This book is an attempt to show just how profound the developments of the last twenty years have been, and how richly rewarding it is to take part, even as a spectator, in the exploration of the planets.

After a brief review of the solar system, the book begins with the Moon, since this is our nearest neighbour in space, and the one about which most is known. It will not be concerned solely with the Apollo missions, but will start by discussing what was known of the Moon before them. This is important, because it will enable the scientific results of the Apollo landings to be seen in a much broader perspective. Some of the remote sensing techniques that have been used on the Moon will also be discussed, because the Moon has provided an excellent testing ground for these techniques before applying them to more distant bodies.

The Moon also provides an ideal starting point for the main function of the book, a review of each of the planets and their attendant satellites. The last chapter attempts to draw all the threads together, to show how all the planets can be considered as a family of related individuals, to establish what elements of common history they may have shared, and how they evolved.

The solar system

Many of the facts and figures about the solar system are so well known, and have been known for such a long time, that it is often difficult to find out how they were first established. There are scores of descriptions of the Sun's family in books, atlases and encyclopedias, and many more in technical works on the subject, yet few of them make any mention of a subject as basic as how to measure the distance of a planet from the Sun. Many books are also crammed with strings of impressive numerical data – the original 'astronomical numbers' – but without any indication of why these numbers are interesting or useful. In this account, it is proposed not only to try and explain how the numbers were arrived at, but also why they are significant. Much of this is rather basic stuff, but a review of it makes the more sophisticated modern data easier to appreciate.

The Sun is much the biggest and the most important member of the solar system. Although worshipped as a god by most civilizations since the beginning of time, it is now known that the Sun is only a star, and an insignificant one at that, located towards the edge of one of the spiral arms of the Milky Way galaxy. Orbiting around the Sun in a well-defined plane are the planets, asteroids, meteorites and comets. Orbiting around some of the planets are miniature solar systems of satellites.

Five of the planets, Mercury, Venus, Mars, Jupiter and Saturn, have been known since men first began to look intelligently at the night sky. Apart from their brightness, they drew attention to themselves by the way that they appeared to wander around against the background of fixed stars: sometimes they drifted steadily in one direction, sometimes they seemed to come to a stop, then move backwards for a while, stop again, and then drift steadily on again. The very word 'planet' comes from the Greek word for 'wanderer'. Naturally, such erratic behaviour made the planets immensely interesting objects, and from very early days they played an essential role in astrology.

The other planets, Uranus, Neptune and Pluto, were not discovered until after the invention of the telescope – Pluto was not discovered until 1930 – and it is possible that there may be one or more planets yet to be discovered. It was once seriously believed that there was a planet between Mercury and the Sun, which would only be visible when it crossed directly across the face of the Sun. Although this idea has long been discounted, there is a slightly better chance that an undiscovered planet exists out beyond Pluto. If it exists, it must be a remote, dim world, extremely difficult to detect even with the most powerful of available telescopes.

The satellites, or moons, circling the planets have in the past been astronomical curiosities, but they are now becoming more and more important in understanding the nature and origin of the solar system. Some of them are large and important objects in their own right. The Earth has only a single – albeit sizeable – attendant. Mars has two tiny ones. Jupiter has no less than fourteen, Saturn eleven, Uranus five, Neptune two, and Pluto one. Pluto's moon was discovered in 1978, while this book was being written. It is probable that spacecraft studies will reveal the presence of other satellites of the outer planets, since these small, distant bodies are near the limits of telescopic observations.

The other members of the solar system are the asteroids, meteorites and comets. There are many tens of thousands of asteroids, but only a handful of these are more than a few tens of kilometres across. The rest are merely irregular boulders, tumbling endlessly through space, most of them on orbits that keep them between Mars and Jupiter, where they form the asteroid belt. In 1977, a small, solitary object named Chiron was found drifting slowly around the Sun between Saturn and Uranus; it may be the first known member of a second great swarm of asteroids.

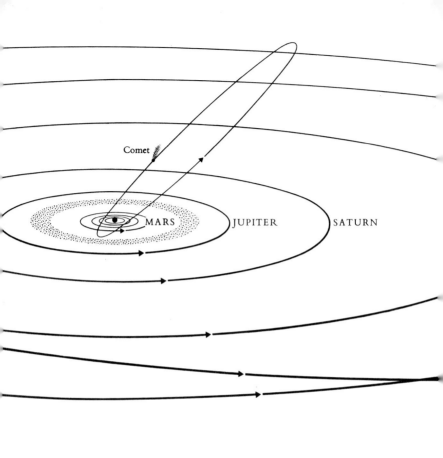

Comet

MARS   JUPITER   SATURN

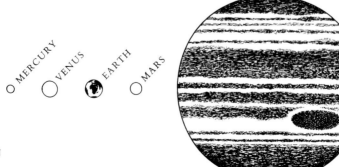

SUN

O MERCURY   VENUS   EARTH   MARS

JUPITER

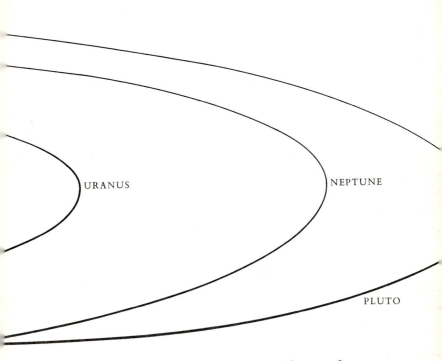

URANUS   NEPTUNE

PLUTO

Figure 1.1. *A summary view of the solar system. The upper figure illustrates the spacing of planets. The orbits of Earth, Venus and Mercury are so small on this scale that they can scarcely be separated. The orbits of all planets except Pluto and Mercury lie roughly in the same plane; Pluto's orbit is markedly inclined to this plane, and that of the comet shown is much more so. In the lower figure, the sizes of planets are compared. The inner planets are insignificant compared to the Sun and the outer planets.*

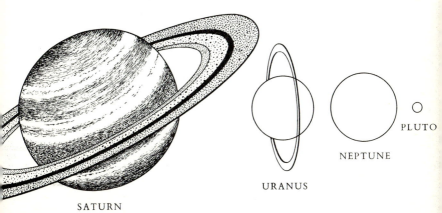

SATURN   URANUS   NEPTUNE   PLUTO

The smallest bodies in the solar system are meteoroids or meteorites, which range in size upwards from small, dust-like grains to boulders as big as an automobile; they may just be broken chunks of larger asteroids.* There is effectively an infinite number of them. Thousands strike the Earth each day, burning up in the atmosphere as shooting stars or *meteors*. Larger ones are comparatively rare, and only a few fist-sized fragments hit the ground each year.

It is not easy to say how many comets the Sun lays claim to, since most of them remain far out on the fringes of the solar system; it is possible that they may be numbered in billions. Although they sometimes blaze spectacularly across the night skies, comets are among the most primitive members of the solar system. They are nothing more than dirty snowballs in space, frigid lumps of ice, frozen gases and dust.

Some statistics

It was easy for the first observers to appreciate that there was something different about the five wandering 'stars', but difficult for them to find out any more about the wanderers, or their relationships to one another and to the Sun. In about 280 B.C., a Greek philosopher, Aristarchus of Samos, deduced correctly that the Sun was the centre of the solar system, and that the planets, including the Earth, revolved around it. His ideas were not, however, accepted by his contemporaries, and so, for the next fourteen centuries, mankind continued to believe that the Earth was the centre of the universe, and that everything else revolved around it. Not until 1512 did a Prussian priest, Canon Nicholaus Copernicus, revive Aristarchus' ideas and postulate a Sun-centred solar system. His work was not published until 1543.

The fourteen-century interval between Aristarchus and Copernicus is one of the most fascinating facets of the whole of the history

* For purists in the use of terminology, it has been suggested that the term 'meteorite' be used only for objects that reach the Earth; while still in space they will be called 'meteoroids'!

of science. Realization of the true nature of the solar system was an essential first step towards Newton's postulation of the laws of gravitation and motion. Although fruitless, it is tantalizing to speculate on what might have happened to the world if the Greeks had accepted a Sun-centred solar system, and also arrived at the laws of gravitation. Would the hydrogen bomb have been discovered by the time of Christ?

Although Copernicus is justly credited with the rediscovery of the Sun-centred solar system, he was still a long way from understanding it fully. He thought that all the heavenly bodies revolved not around the Sun, but about the centre of the Earth's orbit, which was about three Sun diameters from the Sun. Perhaps more important, he still clung to the old Greek notion – usually attributed to the philosopher Plato – that all motion in the universe had to be in perfect circles at uniform speed. To make his theory fit the available observations, he retained another old Greek concept, that of *epicycles* (Figure 1.2), which are small

Figure 1.2. The epicycle concept. Copernicus considered that, while a planet moves around the Sun along a large circular orbit, at the same time it moves round in smaller circles, or epicycles, whose centres lie on the main orbit. The net result of these two movements to an observer on Earth would appear to be as represented by the heavy line: there would be occasions when the planet appeared to be moving backwards.

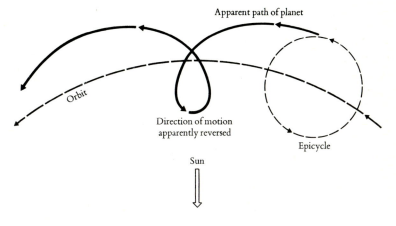

Apparent path of planet

Orbit

Direction of motion
apparently reversed

Epicycle

Sun

circles built up on the bigger circular orbits. In some ways, Copernicus was fortunate, because since there was not much observational data available, he could theorize freely, without having to take note of irritating discrepancies between theory and observation.

The next step forward was made by Tycho Brahe (1546–1601). He, oddly enough, rejected Copernicus' simple idea in favour of a more complex one, in which all the planets revolved around the Sun, but the Sun and Moon revolved around the Earth. Although Tycho was way off the mark in this respect, his work was of the first importance, because he was, above all, an *observer*, and during a lifetime of patient and meticulous observation he increased the accuracy of stellar and planetary data by at least an order of magnitude, without optical instruments. His data were eventually combined together by Johannes Kepler in a complete set known as the *Rudolphine Tables*. These remained in use as a standard reference for over a century, and the planetary data that Tycho had compiled in them enabled Kepler to surmount a great watershed in the course of science: after Kepler, nothing was ever the same again.

Born in Württemburg in 1571, Kepler was an extraordinary character. The son of a noble family, he spent his life in poverty; surrounded by a community totally committed to the idea of an Earth-centred solar system, he was one of the first scientists to speak out in favour of Copernicus' theory; a lifelong believer in a purely mystical, geometrical arrangement of the solar system, he none the less described for the first time the true shape of planetary orbits, and established the three basic laws of planetary motion. Perhaps his greatest contribution to science, however, and the reason why his work marked a watershed, was that he was the first to test his ideas by observation: when he found observations at variance with his hypotheses, he sat down and tried to work out why, and produced a modified hypothesis, rather than merely ignoring them.

On 19 July 1595, Kepler had what he believed to be an inspiration; in a moment of transcendental lucidity, he thought that he had cracked the secret of the solar system. He became convinced

that the spacing of the planets' orbits around the Sun was controlled by the geometry of the five so-called perfect Platonic bodies,
the tetrahedron, cube, octahedron, dodecahedron and icosahedron
(Figure 1.3). At the centre of his solar system lay the spherical

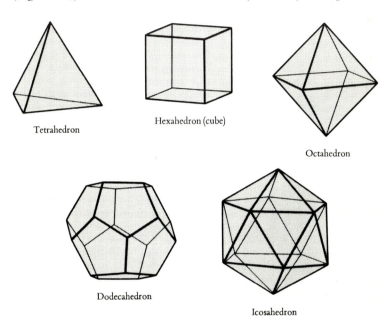

*Figure 1.3.   The five 'perfect' geometric bodies which were at the core
of Kepler's thinking about the nature of the solar system.*

Sun. Snugly fitting around that was the nearly spherical icosahedron, and nestled around that was the sphere that contained the
circular orbit of Mercury. Surrounding that was the dodecahedron, and so on (Figure 1.4).

Kepler was not content with merely establishing this rather
elegant idea as an intellectual concept. He set out to try and *prove*
that it was correct by showing that the ratios of the observed
orbits of the planets coincided with what he predicted. In doing
this, he was as successful as Columbus, who is said to have
discovered America while looking for the Indies.

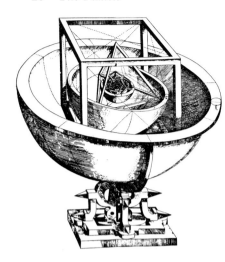

Figure 1.4. Kepler's theoretical model for the solar system. Each of the five geometrical bodies was used to define a spherical shell which contained the orbits of the planets. The outermost shell shown is the one containing Saturn's orbit. (Photo: Anne Ronan Picture Library)

His major task was to measure precisely the radii of the orbits of the planets, which he believed to be perfectly circular. He soon found that he was obtaining results that could not be reconciled with circular orbits, particularly in the case of Mars. He knew that Mars revolved around the Sun every 687 days; thus, every 687 days Mars would be back at a fixed point in its orbit. Using Tycho's observational data, he measured the angle between the Sun, Earth and Mars when Mars was at the fixed point once, and then again when Mars came back to the same point 687 days later. The Earth, of course, would then be in a different point of its own orbit, different by the angle E on Figure 1.5. Knowing these two angles, it is easy to establish the *relative* radii of the Earth's and Mars' orbits. (It is important to note that it was impossible for Kepler to work out the *absolute* radii in terms of numbers of miles or kilometres.)

Kepler's breakthough came when he had repeated this procedure using many different fixed points around Mars' orbit. He found that he could never get the radius of Mars to come out exactly the same twice, and furthermore that, despite the excellent records obtained by Tycho, he could never predict exactly where Mars would be at a given time. He had faith in the observations, however. He considered them to be accurate to about 1 minute of arc;[*] when he found discrepancies of *8* minutes of arc between his predicted positions of Mars and the actual ones, he realized that the concept of circular orbits, which had been rooted in men's minds for millennia, *could not be true.*

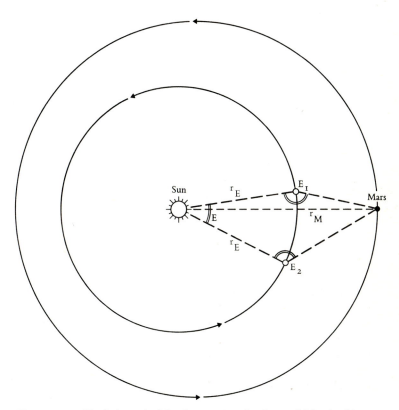

Figure 1.5.   Kepler's method for determining the shape of Mars' orbit.
E_1 *and* E_2 *are positions of the Earth 687 days apart;* r_M *is the radius of Mars' orbit, and* r_E *the radius of the Earth's. Measurement of the two angles* Sun – E_1 – Mars *and* Sun – E_2 – Mars, *and knowledge of the angle E, enabled Kepler to fix Mars' position relative to the Earth.*

[*] There are 60 minutes of arc in 1 degree, and 60 seconds of arc in 1 minute.

After this realization, it did not take him long to deduce that the orbits of the planets are elliptical, not circular, and that Mars' orbit is more sharply elliptical than most. This observation became enshrined in Kepler's First Law: *the planets move in elliptical orbits with the Sun at one focus* (Figure 1.6). Later, he established his Second and Third Laws, which are slightly more complex. The Second Law states that the radius of a planet in its orbit sweeps out equal areas in equal times, and the Third Law that the square of a planet's period is equal to the cube of its mean distance from the Sun. These latter two laws are of great importance, and they led directly to Newton's discovery of the laws of motion and gravitation. Newton was the first to point out the help that Kepler and others had given him. He himself said: 'If I have seen further, it is because I have stood on the shoulders of giants.'

Kepler was working in the days when the telescope was a brand-new invention. There is even in existence a letter written by Kepler to his contemporary, the great Galileo Galilei, pleading to borrow the latter's telescope – Galileo did not even bother to answer. Thus, although Kepler was able to establish the distances of the planets from the Sun relative to the Earth, he could not establish their absolute distances, and he certainly could not say anything about their *sizes*, either relatively or absolutely.

Although the design of telescopes progressed enormously in the centuries after Kepler's death, the determination of absolute distances remained a thorny problem. Since it was much easier to measure distances, as Kepler had done, in terms of multiples of the distance between the Earth and Sun, this distance became formalized as a basic standard of measurement, known as the *astronomical unit*. Thus, if the Earth is taken to be at a mean distance of one astronomical unit from the Sun, Mercury is said to be 0·39 AU distant, and Mars 1·52 AU distant.

Determination of the numerical value of the astronomical unit (the Sun–Earth distance in kilometres) was not really satisfactorily achieved until 1931. The problem is basically simple. When a surveyor is making a map, and wants to measure the distance to some important point a long way off, all he does is establish a

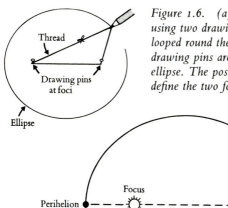

Figure 1.6. (a) An ellipse can be drawn by using two drawing pins and a piece of thread looped round them. The further apart the drawing pins are, the more elliptical is the ellipse. The positions of the drawing pins define the two foci of the ellipse.

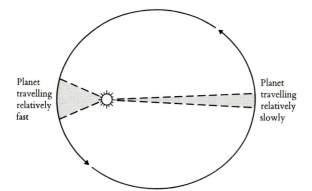

(b) An elliptical planetary orbit. The Sun is at one focus; the other is empty. When a planet reaches the point on its orbit nearest the Sun, it is said to be at perihelion; *when furthest away, at* aphelion.

(c) Kepler's Second Law. The two shaded parts of the ellipse have equal areas. In order for these to be swept out in equal times, the planet has to travel much faster when it is closer to the Sun than when it is further out.

long measured baseline, and then, using a theodolite, measure the angle between his baseline and the object at both ends of his line. He can then calculate the distance to the object by simple trigonometry (Figure 1.7).

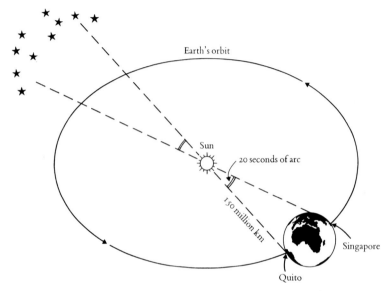

Figure 1.7.   A crude measurement of the distance of the Earth from the Sun can be made by taking bearings on the Sun from two points on the Earth's surface as widely spaced as possible. Knowing the angular difference between the bearings and the distance between the two points, the Sun's distance can be found. Because the Sun appears to move against a background of 'fixed' stars, it will appear to be in a slightly different position from the two observatories. This phenomenon is known as parallax, *and provides a useful method of measurement.*

Things are not so simple when trying to measure the distance of the Earth from the Sun, however. The first problem is that, to get the greatest accuracy, one needs to have as long a baseline as possible; but on Earth one is restricted to the size of the Earth itself. The best one can do is make use of two simultaneous observations as widely spaced as possible, using, for example, one telescope in Quito, Ecuador, and another in Singapore. The next problem is simply that the Sun is *too big*; it is impossible to be sure that the two telescopes are pointing at exactly the same part of the Sun at the same time. To be sure, it is possible to obtain an approximate value for the Sun–Earth distance by this method, but there were wide discrepancies between different sets of results when it was used.

In the seventeenth century, the distances between the Earth and the other planets were well known in terms of astronomical units, and it was realized that, to measure the value of the astronomical unit, it was not necessary to use the Sun itself, because the value could be established by measuring in kilometres *any* distance which was already known in terms of astronomical units. It was also realized that the tiny disc of a planet would provide much better results than the broad face of the Sun.

The parallax method was used with great success in the 1670s by French astronomers observing Mars. Jean Picard and Giovanni Cassini used a telescope at the Paris Observatory to make very precise measurements of the position of Mars relative to the starry background, while one of their colleagues, Jean Richer, made simultaneous observations from an observatory 6,000 kilometres distant at Cayenne in French Guiana, on the northern shores of South America. Richer sweated it out for two years in Cayenne before completing his observations. The major difficulty, of course, was to ensure that the two sets of observations were indeed simultaneous. At that time, precise pendulum clocks had only recently been developed, and Richer was using one to time his observations. A by-product of his work was the discovery that his pendulum beat more slowly than the one in Paris, a fact which was eventually interpreted as being a consequence of the Earth's slight equatorial bulge, and the lower acceleration due to gravity there.

The results of this meticulous piece of long base-line surveying were far reaching. The Sun was found to be 140 million kilometres from the Earth. Although this figure was about 10 million kilometres less than modern work would show, it was an impressive demonstration of the scale of the solar system. For the first time, mankind had a really convincing measure of the vastness of space.

In the following centuries, many other measurements were made, with progressive refinements, but the astronomical unit was not definitively measured until an English astronomer, D. Gill, hit upon the idea of using an asteroid as a kind of stepping stone between the Earth, Sun and planets. The asteroid used was Eros. It occasionally comes within 25 million kilometres of the Earth – close enough for there to be a marked parallax difference – and it is so small that it always appears as a tiny point of light.

In 1930–31, a world-wide concerted effort was made to observe Eros and its parallax. Thirty different telescopes at twenty-four different observatories in fourteen different countries obtained 2,847 photographs in the process and yielded the result that the mean Sun–Earth distance is: 149·6 million kilometres.

With this major step accomplished, it was easy to fill in the rest of the absolute distances for the solar system. These are as given in Table 1.1.

	Million kilometres	*Astronomical units*
Mercury	58	0·39
Venus	108	0·72
Earth	150	1·00
Mars	228	1·52
Jupiter	778	5·20
Saturn	1,427	9·54
Uranus	2,869	19·18
Neptune	4,496	30·06
Pluto	5,900	39·44

Table 1.1.   Mean distances of the planets from the Sun.

Bode's Law

In 1741, when planetary distances were still only known in relative terms, a German astronomer named Wolff realized that there was something rather odd about a table of planetary distances like the one in Table 1.1: the distances are not random, but seem to be governed by some kind of law. This law was developed by another German, David Titius, in 1772, and finally formalized by Johann Bode in 1778. It is now called the Titius–Bode Law, or, more often, Bode's Law. Bode's Law is a bit problematical, and its significance is still not wholly understood, but it played an important part in early studies of the solar system.

Basically, Bode took a sequence of numbers which formed a doubling series:

0   3   6   12   24   48   96   192   384   768 . . .

(But note that there is a hiccup at the beginning of the series – 3 is not double 0.) He then added 4 to each term:

4   7   10   16   28   52   100   196   388   772 . . .

This gives a sequence of numbers which is strongly similar to the planetary distances expressed in astronomical units multiplied by ten (Table 1.2). This may seem to be merely a meaningless

	Mercury	Venus	Earth	Mars		Jupiter	Saturn	Uranus	Neptune	Pluto
Planetary distance (AU × 10)	3·9	7·2	10	15·2	–	52	95	192	300	394
Bode's sequence	4	7	10	16	28	52	100	196	388	772

Table 1.2. Comparison of actual planetary distances with Bode's sequence.

arithmetical coincidence, the result of artificial juggling with numbers. Many astronomers have thought the same. The coincidence cannot be easily ignored, however, because although the 'law' was established in the days before sophisticated telescopes were available, it proved to have remarkable predictive powers.

The first instance of this concerned the obvious 'gap' in the list of planetary distances corresponding to the term 28 in Bode's sequence; there ought to be a planet at a distance of about 2·8 A U from the Sun. A systematic search for the 'missing' planet was begun towards the end of the eighteenth century. This was

Figure 1.8.   The asteroid belt is rather broad, but when it was discovered, Ceres, the biggest of the asteroids, fitted in well with Bode's Law. Many of the other asteroids clearly do not fit nearly as well.

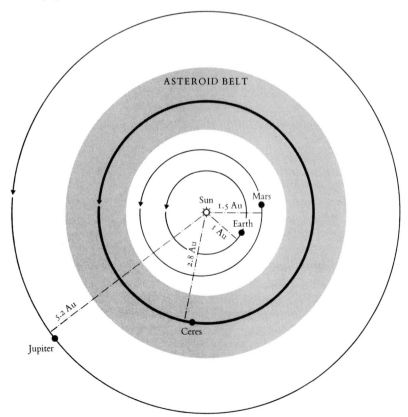

fruitless until New Year's Day 1801, when the Italian astronomer, Giuseppe Piazzi, discovered a very small body which he named Ceres, at a distance of 2·76 AU from the Sun. Although it was too small to constitute the 'missing' planet in its own right, the discovery of Ceres naturally reinforced belief in the validity of Bode's Law, and further observations rapidly led to the discovery of several other small bodies at about the same distance – Pallas in 1802, Juno in 1804 and Vesta in 1807. These are now known to constitute part of the huge swarm of minor planets, or asteroids, that orbit the Sun at distances between 2·2 and 3·2 AU from the Sun.

Another apparent confirmation of Bode's Law came in 1781, when the first of the 'new' planets, Uranus, was discovered by William Herschel. Its distance from the Sun, 19·2 AU, is in

Figure 1.9   An English reflecting telescope of about 1795. This one, made by William Herschel, discoverer of Uranus, is beautifully constructed of wood and brass, but, like many of the telescopes of the period, must have been difficult to use, especially in keeping a planet centred in the field of view. (Photo by courtesy of the Museum of the History of Science, Oxford University)

startlingly close agreement with the 19·6 AU predicted by Bode's Law. Not surprisingly, this close agreement encouraged astronomers to use the law as a starting point in the search for other distant planets. In 1846, the French astronomer, Urbain Le Verrier, and the British astronomer, John Couch Adams, both assumed that a planet was located at 38·8 AU when they set out to try to account for the perturbations observed in Uranus' orbit (which we will return to in Chapter 8). They soon found that this figure would not work, however, and when Neptune was finally discovered, it was located at a distance of only 30 AU. When Pluto was discovered in 1930, the discrepancy between the observed and predicted distances was vastly greater, and it was realized that, despite its past usefulness, Bode's 'Law' does not really dictate the design of the solar system.

Size, mass and density

It is the simplest thing in the world to measure the size of a planet, given a sufficiently powerful telescope and the knowledge of how far away it is. All one has to do is measure the size of the planet's disc in terms of the angle it subtends at the Earth, and then carry out a quick calculation. (The Moon subtends an angle of about 31 minutes of arc, about the same as a pencil held at arm's length.)

Of course, the measurements are never quite as simple as this in practice. For one thing, when measuring a planet with an atmosphere, such as Venus, one cannot be sure of exactly what one is measuring – it might be the solid surface of the planet, or some layer in the atmosphere. For another, most of the planets are not precisely spherical, and most of them, particularly the outer planets, look as though they have been sat upon and squashed, so it is necessary to measure a series of values in order to find the true shape of the planet.

These are relatively minor difficulties, however, and it can be shown that the diameters of the planets are as follows:

Mercury	4,880 km
Venus	12,104 km
Earth	12,756 km
Mars	6,756 km
Jupiter	142,800 km
Saturn	120,000 km
Uranus	51,800 km
Neptune	49,500 km
Pluto	2,700 km

Although there are some impressive numbers in this list, and although the diameters of the planets are of some intrinsic interest, they do not in fact tell us much about the nature of the planets themselves. On size criteria alone, Saturn could be just like the Earth, only ten times bigger. Knowledge of the *density* of the planets is clearly more useful, because it can tell us more about what they are made of. But, of course, it is not possible to measure density directly; it is necessary first of all to establish the *masses* of the planets.

This is surprisingly easy, *provided* that the planet has a moon. All one needs to do is to insert some numbers into a simple equation derived from Kepler's and Newton's Laws, which takes into account the gravitational and centrifugal forces keeping the moon locked into its orbit.★ In this way, the masses of the Earth, Mars, Jupiter, Saturn, Uranus and Neptune can be found.

Mercury and Venus lack satellites, and so, too, did Pluto, until

★ For the mathematically minded, the equation is:

$$M = \frac{4\pi^2 \, d^3}{G \, T^2}$$

where M is the mass of the body required, d the distance of the planet or moon from the central body, G the universal gravitational constant, 6.7×10^{-11} N m^2 kg^{-2}, and T the orbital period of the body. Readers who are fond of number crunching may like to check the mass of the Sun, using the data for the Earth's mean distance and orbital period.

The answer is about 2×10^{30} kg. Make sure you use the right units!

	MERCURY	VENUS	EARTH	MOON
Diameter (kilometres)	4,880	12,104	12,756	3,476
Mass (kilograms)	0.33×10^{24}	4.9×10^{24}	6.0×10^{24}	7.4×10^{22}
Density ($\times 10^3$ kg m^{-3})	5.4	5.2	5.5	3.3

Table 1.3.   Vital statistics of the Moon and Planets.

one was discovered in 1978. Calculating the masses of these planets is much more difficult, and in fact the mass of Pluto remained seriously in doubt until 1978. Even more difficult, of course, is finding the masses of the moons that are so useful in calculating the masses of their parent planets. Very sophisticated computing methods are required to deal with these, so we will steer well clear of them.

Once the sizes and masses of the planets are known, their densities can be rapidly found by simply dividing mass by volume. All the relevant data are tabulated in Table 1.3.★ The data in this table reveal that there are two distinct kinds of planet in the Sun's family. The innermost four are all small and dense; the outermost five are all extremely large and of low density, with the exception of Pluto, which is rather an odd-man-out. The mass of Jupiter is greater than that of the rest of the planets put together, while the density of Saturn is so low that the planet would float on the surface of water, assuming that a cosmic

★ Numerical data are presented in this book in the modern scientific format. To save writing long rows of zeros, large numbers are written as powers of ten: $10^2 = 100$, $10^3 = 1,000$, $10^4 = 10,000$ . . . $10^{26} =$ 100,000,000,000,000,000,000,000,000 and so on. A density written as 3×10^3 kg m^{-3} means 3,000 kilograms per cubic metre. It is also conventional to replace per by a negative index. Thus m^{-1} means the same as per metre, m^{-2} means per square metre, m^{-3} means per cubic metre, and so on.

MARS	JUPITER	SATURN	URANUS	NEPTUNE	PLUTO
6,787	142,800	120,000	51,800	49,500	3,600
$6 \cdot 5 \times 10^{23}$	$1 \cdot 9 \times 10^{27}$	$5 \cdot 7 \times 10^{26}$	$8 \cdot 7 \times 10^{25}$	$1 \cdot 0 \times 10^{26}$	$1 \cdot 6 \times 10^{22}$
$3 \cdot 9$	$1 \cdot 3$	$0 \cdot 7$	$1 \cdot 2$	$1 \cdot 7$	$1 \cdot 5$

hand could pluck it out of the solar system and drop it into a big enough bucket of water.

Because the densities of the inner planets are close to that of the Earth, it follows that they are likely to be composed of broadly similar materials. For this reason, they are often termed the 'terrestrial' planets. The 'giant' planets, by contrast, are quite unlike the Earth, and whereas a great deal can be learned about the terrestrial planets by direct observation and spacecraft landings, the great giants are much more difficult to investigate, and remain shrouded with mystery.

Knowing the density of a planet, however, and its distance from the Sun, makes it possible to speculate sensibly on what it *might* be composed of. The laws of physics also provide some further guidelines. According to the kinetic theory, gases are composed of myriads of moving molecules; the higher the temperature, the faster they move. A gas molecule near the surface of a hot planet such as Mercury (only 58 million kilometres from the Sun) will thus move about very rapidly indeed. Mercury is a small planet with a low mass. Hence, the gravitational force between the planet and the molecule is weak. It follows that it is easy for a fast-moving gas molecule to break away from the gravitational clutches of the planet, and to spiral outwards into the vacuum of space.

For each planet there is, in fact, a critical speed, known as the *escape velocity*, which controls whether or not gas molecules will be retained. The escape velocity is dictated only by the planet's size, mass and surface temperature. If the average speed of the gas molecules exceeds about one fifth of the escape velocity, they will be lost to outer space; if below, they can be retained. So it is clear that a small, hot planet like Mercury is incapable of retaining anything but the most tenuous of atmospheres, while the Earth, substantially larger and cooler, can retain (fortunately for us!) a thin atmosphere of nitrogen and oxygen. The giant planets, by contrast, are not only extremely large, but they are also so far from the Sun that they are exceedingly chilly; hence they can retain thick atmospheres of light gases.

While the concept of escape velocity applies only to the *retention* of atmospheres, similar arguments can be applied to the initial *formation* of planets, as will be outlined in the final chapter. The planets nearest the Sun naturally formed under the highest temperatures, and thus only the least volatile materials, such as iron and rocks, could accumulate; anything else would simply have been sizzled into space. Further out, where temperatures were lower, progressively more volatile materials could accumulate, so the Earth and Mars have much more water than Mercury or Venus, while Uranus, Neptune and Pluto consist almost entirely of frozen gases.

Taking the solar system as a whole, then, one can think of the inner planets as predominantly rocky, with at best only very thin atmospheres, while the outer ones are more like frozen atmospheres with small rocky centres. There are thus two parts to this book. The first, and longest, deals with the inner planets, the ones that we can get to grips with directly. The second deals with the strange giant planets, and what might be concealed beneath the clouds that envelop them. Our knowledge of these planets is in its infancy.

At the time of writing, in 1979, two Voyager spacecraft had just drifted silently past Jupiter, *en route* for a rendezvous with Saturn. They transmitted back to Earth the most visually breathtaking and scientifically informative pictures of a decade which has also seen the first images of the surfaces of Mercury, Venus

Figure 1.10. *One of the best photographs of the Earth ever taken, this Apollo 17 picture encapsulates many of the unique features of the 'good Earth', as the astronauts called it. Well displayed are the Antarctic ice-cap, Madagascar, the Red Sea, and tropical revolving storms.*

and Mars. Such a decade will never recur; it is unique. But one day, perhaps many years from now, a spacecraft will land on one of Jupiter's moons. That will be another great leap forward for mankind, and for science.

2. The Moon

On 25 May 1961, President J. F. Kennedy made a historic declaration in a speech to the Congress of the United States:

'I believe that this nation should commit itself to achieving the goal, before this decade is out, of landing a man on the Moon and returning him safely to Earth. No single space project in this period will be more impressive to mankind, and none will be so difficult or expensive to carry out . . .'

When President Kennedy made this momentous speech, the Russians had already secured photographs of the far side of the Moon, and the Americans were lagging far behind in the space race, as they had been ever since the launch of the first Sputnik in 1957. While it is unfortunate that the Apollo project was conceived primarily for political rather than scientific motives, President Kennedy's commitment of the United States to the project set in motion the most ambitious programme of research and experiment that the world has ever seen, which reached its culmination in the successful Apollo 11 mission. In the ten years before Apollo 11, however, much more was learned about the Moon than in the three hundred years since the invention of the telescope. Before examining the results of the Apollo missions, we shall therefore examine what was known about the Moon before the Apollo programme got under way, and what was learned from the preparatory series of unmanned missions.

Earliest days

Part of the beauty of the Moon for us Earth-bound mortals is that it is always changing. Sometimes it hangs as a slender silver crescent against the evening sky; sometimes it rises as an extraordinarily large ruddy red orb in the east as the Sun sets in the west; sometimes it appears in broad daylight as a pale, washed-out half, lost in the blue of the zenith; sometimes it lights up the midnight landscape in a brilliant silvery radiance. It was natural, therefore, that the Moon should be regarded with awe by almost every primitive civilization, and that Moon worship should be almost universal – it persisted until about the eighth century A.D. in Europe. It may still persist among scattered 'uncivilized' tribes, and among small groups of 'civilized' cranks and eccentrics.

The first people to start thinking seriously about the Moon were the ancient Greeks. Although the classical Greek culture produced some of the most influential philosophers in history, and although the Greeks were outstandingly good at geometry – as we shall see – some of their early ideas on the Moon seem bizarre today.

Anaximander (*c.* 600 B.C.) thought the Moon to be 'a circle nineteen times as large as the Earth; it is like a chariot wheel, the rim of which is hollow and full of fire, as that of the Sun also is, it has one vent, like the nozzle of a pair of bellows, its eclipses depending upon the turnings of the wheel'. Xenophenes (570–478 B.C.) thought the Moon shone with its own light, whereas Heraclitus (540–480 B.C.) envisaged the Moon as a bowl of fire whose changes in attitude caused the visible phases and whose dimness with respect to the Sun was caused by its moving in less pure air. Philolaus, in the late fifth century B.C., thought the Moon to be occupied by animals and plants fifteen times the size of terrestrial equivalents and to have a day fifteen times as long.

With the next generation of Greeks began modern notions of the Moon. Parmenides (512–450 B.C.) was possibly the first to propose that the Earth is round and that the Moon's light is

merely reflected sunlight. Anaxagoras (500–428 B.C.) thought the
Moon to be Earth-like, with plains, mountains and ravines; he
was imprisoned for proposing (correctly) that lunar eclipses are
caused by the Earth passing between the Moon and the Sun. To
Empedocles (493–433 B.C.) is attributed the hypothesis that the
Moon circles the Earth. The Greeks, in fact, probably had a much
better concept of the nature of the Moon than any group up until
the time of Galileo.

Basic statistics: distance, size, mass and density

Because the Apollo missions were such magnificent feats of
technology and human enterprise, it is easy to lose sight of the
wealth of fundamental data on the Moon that was gleaned in
pre-Apollo days, much of it even before the telescope was
invented. Those of us who struggled at school with the tangle of
squares and triangles that is Pythagoras' theorem will be well
aware that it was the ancient Greeks who devised the subject
known as *geometry*. Translated into English, geometry means
something like 'earth measurement'. Not only did they literally
measure the Earth, but the Greeks also made reasonable estimates
of the distances between Earth, Sun and Moon, and estimates of
their sizes.

It was Aristarchus, who first postulated a Sun-centred universe,
who also devised a method for measuring the relative distances of

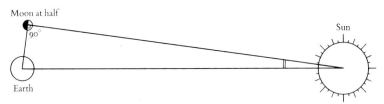

*Figure 2.1.   Aristarchus' method for comparing the relative distances of the
Moon and Sun. When the Earth–Moon–Sun angle is 90°, the angle
Moon–Earth–Sun is also nearly 90°, and is extremely difficult to measure
precisely. Hence, Aristarchus' method could only provide a rough guide.*

Sun and Moon from Earth. He reasoned that when the Moon is half-full, there is a perfect right-angle between the Earth, Moon and Sun. Measurement of the angle Moon–Earth–Sun (Figure 2.1) enables the relative lengths of the sides of the triangle to be compared. Using this method, Aristarchus found that the Sun was much, much further away from the Earth than the Moon – a simple enough conclusion, but a fundamental one, and one never made before. Later, he was able to measure the relative sizes of the Sun, Moon and Earth by measuring the apparent diameters of the Sun and Moon and comparing them with the size of the Earth's shadow cast on the Moon. (This method depended on the fact that it was possible to measure the radius of curvature of the Earth's shadow on the Moon.) Knowing the relative distances

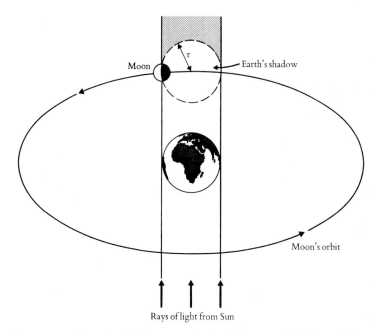

Figure 2.2. Since the Earth casts a circular shadow on the Moon, it is possible to compare their sizes directly, although, of course, the shadow is not so sharply defined as shown here.

concerned, he found that the Sun is vastly bigger than either the Moon or the Earth. It has been suggested that it was his realization of the huge size of the Sun that encouraged him to place it at the centre of his universe.

The next step was to establish the *absolute* sizes of the Sun, Moon and Earth. This was done in the third century B.C. by Eratosthenes, and fortunately there are good records of how he did it. His only instrument was a *skaphe*, which was basically a form of sun-dial, and he used this to make measurements of the shadow cast by the Sun at two different places (Alexandria and Syene) at the same time (Figure 2.3). Using this method, Eratosthenes obtained a value for the Earth's circumference of 39,900 kilometres. This was an amazingly good result, since modern values for the Earth's circumference at the equator are around 40,076 kilometres.

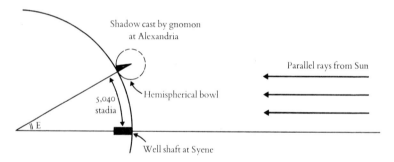

Figure 2.3.   Eratosthenes measured the length of the shadow cast by a vertical pointer at Alexandria while the Sun was vertically overhead at Syene to establish the size of the Earth. The length of the shadow enabled him to find the angle of the Sun from the zenith at Alexandria, which must be the same as the angle E. Knowing the distance between the two towns, he could then calculate a value for the radius of the Earth.

Eratosthenes' data, coupled with that of Aristarchus, thus provided the means of obtaining the first solid piece of *quantitative* data about the Moon: an estimate of its size. Hipparchus (*c.* 162–126 B.C.), using Eratosthenes' figures, calculated that the Moon was about one third the size of the Earth; this figure was refined later by Ptolemy to about 0·29 of the Earth's diameter. The correct value is 0·272 Earth diameters, corresponding to 3,476 kilometres.

Ptolemy also used Eratosthenes' data to establish a good value for the next most important statistic, the Earth–Moon distance. When the Moon was directly overhead at one point (which might have been Syene again), the angle of the Moon from the zenith was measured at Alexandria (angle ZAM on Figure 2.4).

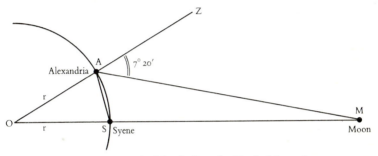

Figure 2.4. Ptolemy's method for finding the Earth–Moon distance.

Knowing the distance between the two towns and the radius of the Earth, r, the Earth–Moon distance OM can be found by geometry. Ptolemy obtained a value of about 355,000 kilometres. This was also a good estimate, since modern values show that the Moon at its nearest is 356,000 kilometres distant, and at its furthest 406,000 kilometres. It is a great pity that we do not know what instruments Ptolemy used to obtain such accuracy. It is known, however, that navigational instruments of impressive accuracy were in use on Greek ships at the time, so he probably used something similar.

Apart from this useful piece of geometry, there was one other important fact about the Moon that the Greeks – and every other civilization before and since – could deduce: it always keeps the same face towards the Earth. This is the first example of an elegant phenomenon that we will meet more than once in our exploration of the solar system. The phenomenon is called *spin-orbit coupling*, or *resonance*. In the case of the Moon, it means simply that the Moon spins round once on its own axis in *exactly* the same time as it takes to complete one orbit around the Earth. This period is 27·3 days, and is called the *sidereal* month. The more familiar lunar month, which is the time taken for the Moon to go from new moon to full and back again, is the *synodic* month, and lasts 29·5 days. (Neither of these, of course, is the same as a *calendar* month.) The difference between the two is simply that, since the Moon is both spinning on its axis and moving round its

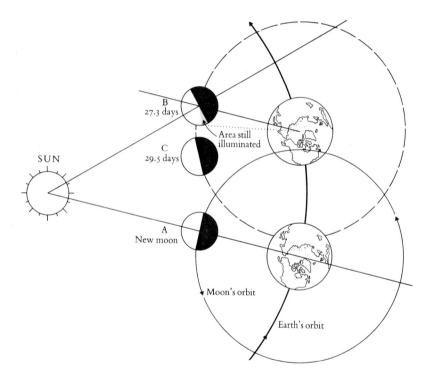

Figure 2.5.   The synodic and sidereal months. In one sidereal month (27·3 days), the Moon has completed one orbit around the Earth, but has moved in space from A to B. In order to appear *again as a new moon seen from Earth, it has to continue to C, which it reaches in 29·5 days, the synodic month.*

orbit, it has to spin through more than 360 degrees in order to appear in an equivalent position as seen from the Earth.

The fact that spin–orbit coupling keeps one side of the Moon permanently facing the Earth has given rise to all kinds of legends and folklore about the 'dark' side of the Moon. While successive space missions have shown that the dark side is not dark, and not very mysterious, they have shown that the coupling is the result of the uneven distribution of mass within the Moon. The gravitational effects of the Earth act on this asymmetry, and in effect provide a braking force which keeps the more massive part of the Moon facing the Earth.

But how does one find out what the mass of the Moon is, let alone how it is distributed? The mass of the Moon is something we must examine here, because it is perhaps the single most important statistic in establishing what the Moon is made of. The ancient Greeks worked out the approximate size and distance of the Moon; but it was many centuries before the mass could be found. As discussed in Chapter 1 (page 31), it is easy to calculate the mass of a body if it has a satellite orbiting round it. The Moon, alas, has no satellites and therefore other indirect methods had to be used.

The first scientist to address himself to this difficult task was Sir Isaac Newton. He tackled the problem by considering the gravitational effects of the Moon on the Earth. The simplest expression of these is in the rise and fall of the ocean tides: the larger the mass of the Moon, the higher the tidal range would be. There are many dreadful complications in this simple statement, most important of which are those introduced by the tidal effects caused by the Sun's mass. Newton observed that tides are highest at new moon, when Sun and Moon are pulling together, and lowest at full, when they are opposing one another. In this way, he sought to compare the masses of the Sun and Moon, but his result was rather inaccurate. He calculated the mass of the Moon to be about one thirtieth that of the Earth.

The French scientist, Pierre Simon de Laplace, later took up the problem. By measuring the tidal ranges at Brest, he came up with a value of about one sixtieth the mass of the Earth. The wide discrepancy between his and Newton's results made Laplace consider the whole technique unsatisfactory, so he devised other more sophisticated methods based on astronomical parameters such as the minute distortions in the Moon's orbit caused by the Earth's shape not being perfectly spherical. Laplace finally concluded that the mass of the Moon is about $\frac{1}{71}$ that of the Earth, equivalent to about $8\cdot3 \times 10^{22}$ kg. The modern value is quite close: $7\cdot4 \times 10^{22}$ kg.

This discussion of the calculation of the Moon's mass may seem to be esoteric to a high degree; an intellectual exercise for withered, academic astronomers working away in decaying ivory towers. Not so. Knowing the size and mass of the Moon, we can immediately calculate its *density*; it turns out to be about $3\cdot3 \times 10^3$ kg m^{-3}.

If this still appears to be just another number, remember that the density of green cheese is about $1\cdot2 \times 10^3$ kg m^{-3}: we have just proved that the Moon is not made of green cheese! More seriously, since we know that the density of the Earth is about $5\cdot5 \times 10^3$ kg m^{-3}, it is clear that there must be major differences between the compositions of the Moon and Earth. Since the most massive part of the Earth is its metallic core (density about 10×10^3 kg m^{-3}), it is a reasonable first guess that the Moon lacks such a dense core.

One other important statistic before we get down to the details of what can actually be seen on the Moon: knowing the size and mass of the Moon, we can also calculate the *acceleration due to gravity* at the surface. It is about $1\cdot7$ m sec^{-2}, only about one sixth of that at the Earth's surface. This interesting little fact has been much used in science-fiction stories about what it would be like to live and work on the Moon. The lunar explorer would be able to bound over obstacles like a kangaroo, hurl cricket balls for kilometres and lift heavy weights that would leave him gasping on Earth. Such stories create the impression that living on the Moon would be like living perpetually in a kind of Olympic stadium.

There are some much more important consequences of the lower lunar gravity, however. For example, a volcano erupting on the Moon would hurl ash particles higher and further than on

Earth. Similarly, the way in which loose material moves down slopes on lunar hills is different from that on Earth. Such effects help to account for the gently rolling, softly moulded lunar landscape that is so startlingly different from the jagged, hostile terrain portrayed by early science-fiction writers.

Figure 2.6.   The full moon, seen through one of the world's greatest telescopes at Mount Palomar in California. Note the contrasts between the pale, heavily cratered highlands and the dark, smooth plains or 'seas', usually called by their Latin name, maria. *(Hale Observatory photograph)*

Several apparently irregular grey splodges can be seen on the Moon's surface with the naked eye. These grey patches have been identified with all kinds of strange men and animals, evidence of the universal tendency for observers to romanticize their observations and to try and make more of them than is factually warranted. Man did not get a true idea of the nature of the lunar landscape until the first spacecraft sent close-up pictures. Before then, he was confined to what he could see with telescopes. But it is surprising how many astronomers did not limit their interpretations to what they could see through their telescopes: they consistently let their imaginations run well ahead of what it was physically possible to see, and this led to some serious misinterpretations, some of which are discussed in the present chapter.

The telescope was probably invented in 1608. By the summer of 1609, Thomas Harriot, an Englishman, had used an instrument to make the first known observations of the Moon, and later the same year Galileo in Italy made some observations, and published the first lunar maps. These look a bit crude (Figure 2.7), but it is possible to identify some individual features on them. Describing his observations in a publication known as the *Siderius Nuncius* ('Messenger from the Stars'), Galileo wrote:

> The surface of the Moon is not perfectly smooth, free from inequalities and exactly spherical, as a large school of philosophers considers . . . on the contrary it is full of irregularities, uneven, full of hollows and protuberances, just like the surface of the Earth itself, which is varied everywhere by lofty mountains and deep valleys.

This passage was particularly important because it helped to dispel finally the long-held belief that everything in the universe had to be smooth and spherical and move in uniform circles at uniform speed. This was the basis of the Platonic school of philosophy, which dated back for hundreds of years. Copernicus had begun to dismantle this philosophy with his work on the motions of the planets around the Sun, but the deeply conservative religious authorities resisted his ideas. Galileo himself got into trouble with the authorities for siding with Copernicus, and was forced to recant some of his statements. His early work on the

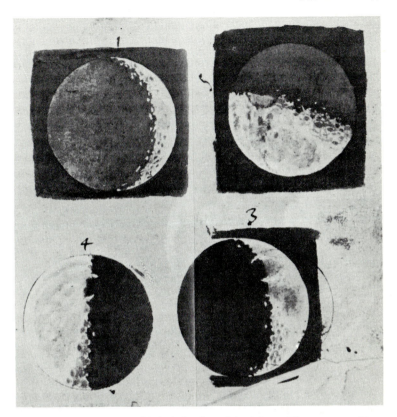

Figure 2.7. *The first published sketches of the Moon. These appeared in Galileo's* Siderius Nuncius *('Messenger from the Stars'). The woodcuts show craters clearly, and Galileo correctly deduced from the fact that the boundary between sunlit and dark sides of the Moon is highly irregular that the Moon's surface is rugged and mountainous.*

Moon, however, remains as evidence of his originality and excellent observation.

In the centuries after Galileo, the quality of telescopes improved enormously, and more and more astronomers worked on the task of building up maps of the surface features. The first map of the Moon to have features systematically named was that of Langrenus in 1645. Langrenus was a scientist in the court of Philip II of Spain. He seems to have been a dreadful sycophant, because he named everything he saw on the Moon after members of the Spanish aristocracy; almost anyone with whom he wanted to curry favour had a crater named after them. His system never became established, though a crater named after him is still there!

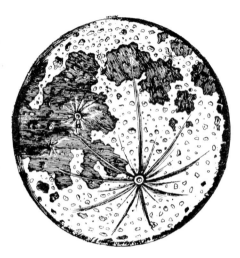

Figure 2.8.   A drawing of the Moon by Fontana in 1629, one of the best early telescopic sketches. The outlines of the 'seas' are recognizable, and the rayed crater Tycho is unmistakable.

The system that did eventually come to be adopted was devised by an Italian astronomer, Giovanni Riccioli, in 1651. He thought that the dark areas of the Moon, which make up the face of the 'Man in the Moon', were seas, so he called them all by various watery Latin titles, ranging from Mare Fecunditatis (Sea of Fertility), through Oceanus Procellarum (Ocean of Storms) to Lacus Mortis (Lake of Death). All of these names are still in use today, but, strangely, the names that Riccioli gave to the light-coloured 'highland' parts of the Moon have been forgotten. This is odd, because the names he used were much the same in style as those he gave to the seas – Land of Heat, Land of Health, Land of Sterility and so on.

For the lunar craters, Riccioli used a different method. He named each one after a famous astronomer, starting with the early

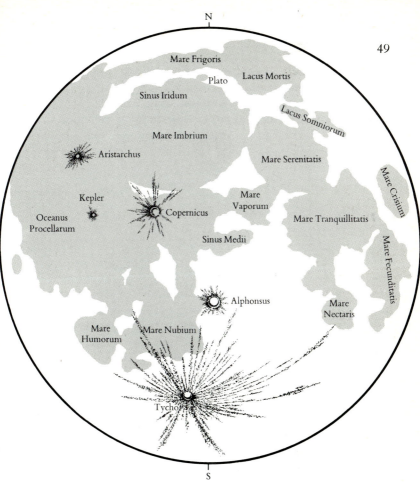

Figure 2.9.   An outline map of the Moon. Only the principal 'seas' and a few of the most important craters are shown.

Greeks, such as Archimedes and Eratosthenes, and went on to include his near contemporaries, such as Copernicus and Kepler. Subsequently, this system was universally accepted by astronomers, who, as they discovered more and more craters, simply named them after each other.

As telescopes grew bigger and better, and the sizes of discernible craters smaller and smaller, the Riccioli system became unwieldy, and a number of inconsistencies crept in. René Descartes, the great mathematician-philosopher, has a prominent crater near the centre of the Moon named after him, but his contemporary Newton has to be content with an insignificant little hollow tucked away near the south pole. Even Galileo is commemorated by a thoroughly obscure little crater right at the edge of the Moon!

Motions of the Moon

It may seem superficially that the Moon moves in a simple way: it rises in the east, sets in the west and appears to move steadily across the sky. The true situation is far more complicated, and with the aid of the telescope it is possible to make precise measurements of the shape of the Moon's orbit and its velocity at different points around it. Surprisingly enough, such painstaking measurements had a direct practical use on Earth at one time. Before the invention of the chronometer in the eighteenth century, navigators had been baffled by the task of determining their longitude on the Earth's surface. They could determine latitudes easily enough, but the problem of how to measure longitude defeated many of the best brains in Europe. Because of its importance to mariners, many different methods were explored, and the British government offered a large cash prize as an inducement to innovation. One of the few methods to become established before the problem was definitively solved by the arrival of the chronometer involved elaborate measurements of the Moon's position. The method was slow and clumsy, but it was all that many of the great navigators of the eighteenth century possessed.

More recently, it would clearly have been impossible to carry out *any* kind of spacecraft exploration of the Moon without being able to predict its position in space at any given instant to an extremely high degree of accuracy. The mathematics involved in working out the details of lunar motions are formidable, not only in the level of sophistication but also in the sheer number of calculations required. Wallace J. Eckert, who worked on the subject from 1952 to 1971 during the period leading up to the Apollo project, used 6,000 terms in his solution!

An aspect of lunar motion that is rather easier to deal with is the phenomenon of *libration*. Since the plane of the Moon's orbit is inclined at an angle to the plane of the Earth's orbit, called the *ecliptic plane*, it is possible to see different parts of the polar regions of the Moon at different times during its orbit around the Earth. Figure 2.10 shows that the northern polar regions are visible when the Moon is below the ecliptic, whereas the southern polar regions are visible when it is above. This phenomenon is known as *libration in latitude*.

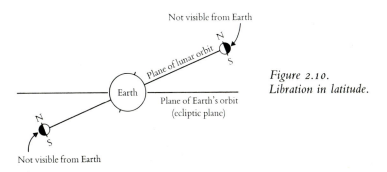

Figure 2.10.
Libration in latitude.

Although it was stated on page 41 that the Moon exhibits spin-orbiting coupling, which keeps one face always towards the Earth, the situation is a little more complex. In accordance with Kepler's Second Law, the Moon's velocity in its orbit varies slightly from one extreme of its orbit to another. The rate at which it spins on its axis remains constant, however, and this means that the orbital period is sometimes slightly ahead of, and sometimes slightly behind, the axial period. This enables us to catch a glimpse of features just beyond the eastern and western edges of the Moon, and is known as *libration in longitude*. The net result of both forms of libration is that we get a chance to see a little over the top and bottom of the Moon, and around its sides. Thus about 60 per cent of its surface is visible from Earth, instead of a simple 50 per cent.

Lunar photography

The first photographs of the Moon were taken in 1840, only one year after the development of the first successful photographic method, the Daguerreotype process. The potential value of photography was realized immediately, and a French scientist, Arago, claimed in 1840: 'We shall be able to accomplish one of the hardest tasks in astronomy – mapping the Moon – in a few minutes.' This was not the case.

Lunar photography has both advantages and disadvantages compared with visual observation. The major disadvantage is that the eye can discern through a telescope details which are about four times smaller than those which can be photographed through the same telescope. This is a consequence of the exposure time needed to take a photograph, and of the ability of the human eye to take advantage of momentary variations in the 'seeing' conditions produced by disturbances in the Earth's atmosphere.

The major advantage is that a photograph is a fixed, permanent record. It can be used for comparison purposes months or years after it was first taken, and it is also much easier to make measurements on a photograph than through a telescope, which necessarily involves working late at night in a chilly observatory. Even more important is the question of objectivity: any drawing of a visual observation is a highly subjective record, no matter how conscientious the observer.

Unfortunately, a tradition of visual observations coupled with line drawings developed during the latter part of the nineteenth century and the first part of the twentieth century, and this has

Figure 2.11.   This photograph of Copernicus crater and the surrounding area was taken with a giant 100-inch telescope at Mount Palomar in California. Copernicus is about 80 kilometres in diameter, but the smallest feature that can be seen on the Moon with any telescope is about half a kilometre across. An Olympic-sized athletic stadium might just be visible. It would be an inconspicuous spot on the floor of Copernicus. (Hale Observatory photograph)

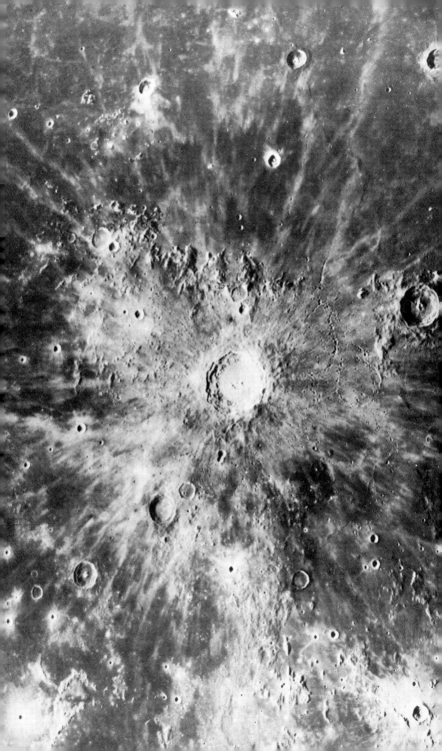

been held responsible for geologists showing little interest in the Moon during this period; study of the Moon was the special province of a small clique of observational astronomers. Although the quality and size of telescopes steadily increased over this same period, it could never be argued that knowledge of the Moon increased at anything like the same rate. For several decades after the First World War, lunar science seems to have been almost stagnant. One of the surprising facets of this period of stagnation is that many of the best lunar photographs ever taken were made before 1920. These were not excelled until the preparatory work for the Apollo missions commenced, and are still reproduced in many textbooks.

Measuring the Moon

Almost as soon as astronomers began using telescopes on the Moon, they began to document their observations with measurements of the heights of mountains and depths of craters. Even Galileo attempted crude measurements of this kind.

Relative heights are the simplest to determine. They are no more than the heights of objects, such as mountains, above their surroundings, and they can be determined by measuring the length of the shadow cast by the object. Knowing the angle of elevation of the Sun at the time, it only requires trigonometry to find the height of the object. Such measurements enable topographic maps of the Moon's surface to be prepared. The highest points rise about 8,000 metres above their surroundings, comparable, say, with the height of Everest above the Indian plains.

Accumulated measurements of this kind enable contour charts of the Moon to be drawn up. It is clear that the lowest parts of the Moon's surface correspond to the grey splodges making up the Man in the Moon. The paler-coloured areas are all highlands. The differences between these two types of lunar terrain are of first importance and will crop up several times in subsequent sections.

Figure 2.12. A contour map of the Moon's surface.
Compare this with Figure 2.9 – the 'seas' are all low-lying areas.

More sophisticated measurements are required to work out the exact shape, or *figure*, of the Moon. It is easy enough to observe the profile of the Moon when seen against the starry background, but it is also necessary to establish its profile in the direction at right-angles to this. By painstaking observation of the heights of points on the lunar surface, it was possible to show that the Moon – like the Earth – is not a true sphere, but is slightly flattened and has a small bulge facing towards the Earth.

More important than this is the fact that the Moon's *centre of mass* is offset from its centre of figure by about 2 kilometres (Figure 2.13). This means that the Moon's 'centre of gravity' is not at its geometric centre – where you'd expect it to be – but is displaced slightly towards the Earth, and that the crust of the Moon is in effect thinner on one side than on the other. This

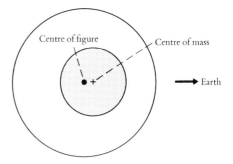

Figure 2.13.   The Moon's centre of mass is offset by about 2 kilometres from its centre of figure, or geometric centre.

inhomogeneity in the Moon is what is responsible for its keeping one face always towards us. If it tried to turn away from us, it would have to try and move its centre of mass away; and to do this would require work to be done. The principle is the same as that of an unbalanced automobile wheel. If freely suspended on its axle, clear of the ground, the wheel will always rotate until the heaviest part of it is vertically below the centre. The resonance between the Moon's orbital and axial rotation periods is an expression of the fact that the Moon has reached a situation of energy equilibrium. To rotate on its axis at any speed other than its present one would require extra energy to be derived from somewhere.

The Earth–Moon distance: stop the world, I want to get off!

It does not require great insight to perceive that the gravitational interactions between the Earth and Moon are affected by the distance between the two bodies, or to realize that this distance may vary through time. But who would dream that the means of finding out how the distance has varied could be found in the shells of long-dead sea creatures?

The link lies in the number of days in a lunar month, the number of days in a year, and the number of hours in a day. If this still seems obscure, recall that almost all animals – humans included – are affected by diurnal rhythms of one kind or another. Such rhythms leave no visible mark on humans, although there are unmistakable signs when the rhythm is disturbed – 'jet lag' is an example of this. Some marine animals leave a permanent record of the passage of night and day in the construction of their shells. Much the best example is the nautilus, the extremely beautiful cephalopod that lives in the south-western Pacific ocean. Nautilus shells are often displayed in museums, cut in half to show the elegant spiral of internal chambers.

Now nautilus characteristically spends the day in deep water, and comes up towards the surface at night to feed. This regular cycle is such an entrenched part of the animal's metabolism that it leaves a clearly defined growth increment on the outer part of its living chamber – the growing edge of its shell, so to speak. Marine biologists have studied living nautilus specimens, and found that every thirty days they secrete a wall of shell at the end of their living chamber, forming a new internal chamber. Since nautilus is unmistakably a nocturnal animal, and since the number of growth lines per chamber (thirty) is so close to the lunar synodic month (29·53 days), it seems likely that it is the length of the month that controls the number of growth lines per chamber. This conclusion is borne out by the presence of similar periodic structures in the shells of other marine organisms.

So far so good: living nautilus shells are governed by the length of the lunar month. But cephalopod species almost identical to nautilus have existed on Earth for hundreds of millions of years, and they are really rather common as fossils. This is where the real interest lies. Examination of perfectly preserved fossil nautiloids 420 millions of years old reveals that these have only *nine* growth lines per chamber, and that in younger fossils the number gradually increases up to the thirty of the present day.

Thus, the number of days in a lunar month has been steadily increasing through geological time.

The implications of this are startling. First, the Earth's period of rotation must have been very much shorter 400 million years ago than it is now – it must have been spinning much faster. Secondly, there must therefore have been many more days in the Earth's year. Thirdly, the Moon must have been much closer to the Earth than it is now, perhaps only half the distance. Fourthly, and finally, the Moon must have revolved around the Earth much more quickly than it does today.

These four factors may appear to be completely unrelated, but they are, in fact, inseparably tied together, because they involve the principle of conservation of energy. The Moon keeps the same face towards the Earth because it would have to do work to turn its centre of mass away. The effect of the Moon on the Earth is to raise large tidal bulges in the oceans. These bulges themselves interact gravitationally on the Moon, causing its axial rotation to slow down and to increase the size of its orbit by moving away from the Earth. The tiny amount of work done on the Earth by the movement of the water masses against friction simultaneously causes the Earth's rotation to slow down.

This may all seem pretty tenuous, based, as it is, on observations of fossil shellfish. There are, however, excellent theoretical reasons for believing that the Earth–Moon system has been evolving in this way through geological time, and also some direct observational evidence. This comes from historical records of lunar eclipses, of which there are fortunately usable records going back nearly 3,000 years. The historical records suggest that the Moon is drifting away from the Earth at 6 centimetres a year, and that the Earth's period of rotation is increasing at the rate of 25 seconds per million years.

The data from fossil nautiloids suggest a much faster rate of recession for the Moon over the last 400 million years, and a much greater lengthening in the Earth's rotation period. The discrepancy between the two sets of data is yet to be reconciled – it would not be surprising if some of the early historical records were inaccurate. It is inescapable, however, that the Moon is drifting away from us, and that the Earth is slowing down. It will be some time before it stops, unfortunately, so none of us will be able to get off in this lifetime!

Using the Moon's light: the electromagnetic spectrum

Measurements of sizes and distances on the Moon are fairly straightforward tasks. But how does one set about finding out more about the Moon, given only a telescope? Can one say anything about what it is made of? How hot it is? To answer questions like these, one has to dissect the light from the Moon as minutely as a zoologist dissects an animal.

The cold, silvery radiance of moonlight that gives such an ethereal quality to the night-time landscape and forms a shimmering, shifting track on the surface of water is nothing more than second-hand sunlight, reflected by the surface of the Moon.

Figure 2.14. The chief components of the electro-magnetic spectrum. The visible portion forms only a tiny fraction of the whole. Almost all wavelengths shorter than the visible are blocked by the Earth's atmosphere.

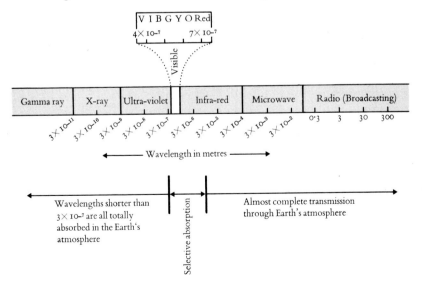

Two things happen to the light after it leaves the Sun. First, part of it is absorbed by the Moon itself, and second, part of it is absorbed by the Earth's atmosphere. What we refer to as 'sunlight' is, in fact, a continuous spectrum of radiation, of which the visible colours form only a part. Each part of the spectrum is characterized by its own wavelength, and different parts of the spectrum are absorbed or reflected differently in different conditions. It is these variations that astronomers use to study the surfaces of the Moon and other bodies.

Photometry

The simplest quantity that can be measured with the Moon's light is how bright it is. This can be done by using an instrument like an ordinary camera exposure meter, and makes it possible to measure the Moon's *albedo*, the ratio of the light falling *on* the Moon to that reflected *by* it. Such measurements show that, although the Moon appears as a magnificently bright body, dominating the night sky, it is really rather dark, and reflects only about 7 per cent of the sunlight falling on it. This is comparable with the reflectivity of exceedingly dull, grey-coloured terrestrial materials with matt textures, such as ash. So here is the first clue to what the surface of the Moon is really like: although the soft gleaming light of the Moon has great romantic appeal, it is really rather a drab place.

One can go further by exploring the reflectivity of the Moon in different parts of the spectrum. This is the technique known as *spectrophotometry*. Through a powerful telescope, the Moon appears as a rather uniform-coloured object; this indicates that it absorbs most wavelengths uniformly. Some observers, however, have recorded slight variations in colour of various features, such as the floor of individual craters. Instruments have also been used to measure the intensity of light from the Moon at different wavelengths, and these have been compared with the figures obtained from samples of terrestrial materials. Such studies showed that the light from the Moon is comparable with that reflected by coarse-grained volcanic rocks.

Polarimetry

When light is reflected from a surface – any surface – it becomes partially polarized. That is why 'Polaroid' sunglasses are so valuable; they cut down the reflected glare from shiny surfaces such as the sea and automobile bonnets. The amount of polarization depends on the angle at which the light strikes the surface, and on the nature of the surface. In general, light-coloured surfaces are less polarizing than dark ones, and matt-textured surfaces are less polarizing than shiny ones.

Studies of polarization of the Moon's light showed that it is much more strongly polarized when it is waning than when it is waxing. This is because, when the Moon is waxing, there is a much larger proportion of light 'highland' material being illuminated than when it is waning. Virtually the whole of the western hemisphere is composed of dark grey patches making up the great 'seas' like the Mare Imbrium.

Polarimetry may not seem at first sight to be a particularly impressive technique, but it is a powerful tool in remote sensing. The French astronomer, Audoin Dollfus, has, in particular, championed its use. He measured the variations in degree of polarization of the lunar surface at different angles of illumination and then examined a wide range of terrestrial materials to find those which gave the closest approximation to the lunar data.

After many trials, he found that the best match came from powdered volcanic lavas, particularly *basalt* lavas of the type erupted by most terrestrial volcanoes, such as Etna and Stromboli. To give the right characteristics, however, the volcanic dust not only needs to be opaque and fine-grained, but the grains also have to be stacked in a distinct structure. This has become known as 'fairy castle' structure. To anticipate later sections somewhat, it was found that when lunar dust samples were returned to

Earth, they would only yield the same polarization properties if this complicated grain structure was preserved; when the same dust was pressed flat, the properties were significantly different.

The 'fairy castle' structure means that it is much easier for light to get out in its incident direction than any other. This also explains why it is that the shadows of objects on the lunar surface – such as astronauts – which are directly down Sun from the camera, always appear on photographs to be surrounded by bright haloes.

Polarimetric studies therefore provided pre-Apollo astronomers with a good guide to the nature of the Moon's surface: it is covered with dust, the dust being probably of basaltic composition and fine-grained in texture. The origin of the dust was not difficult to explain: it must be the result of the countless thousands of meteorite and micrometeorite impacts on the surface. Some scientists went much further, and an American, Professor Tom Gold, even went so far as to suggest that the surface was covered in immensely thick drifts of dust, making spacecraft landings impossible. Such suggestions had to be taken seriously during the lead-up to the Apollo period, and led to some excellent science-fiction stories about spacecraft being swallowed up in seas of dust. Fortunately, these gloomy speculations were not substantiated by later studies.

Taking the temperature of the Moon

Lying on the beach on a fine sunny day, we can feel the heat from the Sun warming us. It is not too difficult to work back to find out how hot the Sun's surface is. Lying on the beach on a fine moony night can be a chilly experience, but a sensitive instrument could detect some heat, and again it is possible to work back to find the surface temperature of the Moon. The situation is complicated, however, because the heat that is measured comes from three sources. A large part of it is simply reflected solar radiation. The second is solar radiation which is absorbed by the lunar surface, and then re-radiated. Finally, there is a tiny component arising from the Moon's internal heat.

As with visible light, studies of thermal radiation from the Moon give different results at different wavelengths: short wavelengths provide data about the material at the extreme surface of the Moon, longer wavelengths about slightly deeper material. All the data show that the Moon is an extremely chilly place. The short wavelength studies revealed a temperature maximum of 87 °C during the lunar day, and a minimum of − 120 °C at night, whereas the longer wavelength studies revealed a more or less constant temperature of − 83 °C, which must represent the equilibrium between incoming solar heat, re-radiated solar heat, and the Moon's internal heat.

Radar

Most people these days are aware of the enormous contribution that radio astronomy has played in our understanding of the most distant parts of the universe – galaxies, quasars and the rest. The mere size and shape of a giant radio telescope suggest a quest for something remote and mysterious. Who can fail to be inspired by the sight of the great bowl of a radio telescope pointing up at the heavens, seeking answers to questions that most of us would find hard to formulate; questions that are among the most fundamental that Man has tried to answer?

The contribution that radio astronomy, and radar in particular, has made to our understanding of the Moon and planets has been of comparable value, but is less widely appreciated. This is probably because the data are more difficult to understand, and do not have the immediate appeal of pictures of the planets taken with optical telescopes.

Just after the Second World War, in 1946, workers in Hungary and at the Army Signal Laboratory in the United States succeeded in detecting radar echoes from the Moon. This was a milestone in lunar science. It was exciting for two reasons.

Previously, to find out anything about the Moon, astronomers had been forced to rely on whatever natural radiation happened to be available *from* the Moon, rather a passive form of research.

When radar became available, they could work on a much more *active* basis, and beam at the Moon quite different kinds of radiation, which they could control at will. Furthermore, while ordinary optical telescopic techniques provided data only about the extreme outer surface of the Moon, the radar echoes bounced back from the Moon provided a method of digging just a little deeper.

The technical details of radar work are likely to be of interest only to radio enthusiasts; the results obtained are more important here. Radar can be used on the Moon in two basically different ways: either to examine the properties of the whole surface of the Moon, or to look at individual small patches about 500 metres across. The 'whole Moon' studies were useful in building up a concept of what the 'average' lunar surface is like. They showed that the entire surface of the Moon is covered with small craterlets one or two metres in diameter. These are quite invisible to Earth-based telescopes, which are incapable of making out details less than about 500 metres in diameter. They also showed that a large number of buried boulders are scattered around everywhere.

The work on the small 500-metre patches was used to build up radar maps of the Moon. These studies confirmed that the surface everywhere is loose and rubbly, as had been predicted by earlier optical work, and that it has similar properties to a dry, dusty desert, with the average grain size of the surface materials being around one hundredth of a centimetre. Comparisons with radar data on ordinary terrestrial rock showed that basaltic volcanic rocks gave the best fit to the lunar data, in agreement with optical work.

Radar and the other telescopic techniques discussed in this chapter are, of course, not now particularly important in lunar studies – men have been to the Moon and brought back samples of surface material, and have collected huge volumes of data while actually on the Moon. It is, however, enormously important to know that these techniques exist. Their value has, if anything, been enhanced by the Apollo missions, because their results can be directly cross-checked with direct evidence from the Moon, and this means that they can be applied with greater confidence to *other* planets and their satellites.

Radar in particular has completely changed our view of Venus, and has provided major advances in our knowledge of Mars and Mercury. It will be many years before samples are obtained from the surfaces of other planets, so these remote sensing techniques will remain extremely valuable. Almost our only knowledge of the nature of the asteroids comes from humble photometric and polarimetric techniques, and this situation is likely to continue until spacecraft can be landed on them.

Surface features of the Moon: the telescopic view

So far this chapter has kept strictly to the straight and narrow paths of lunar observation, and steered away from the rather more risky business of dealing with how the Moon's surface features are to be interpreted. With the benefit of hindsight, and billions of dollars' worth of direct investigation, many of the ideas of the telescopic astronomers now seem amazingly bizarre. It is perhaps not surprising that the invention of the telescope did not *immediately* dispel the notion that the grey areas on the Moon were true 'seas', and that the Moon was thickly populated by intelligent races. It is much more surprising that, in the 1920s, one distinguished astronomer was still talking of certain dark patches on the Moon as being due to migrating herds of lunar animals, and that, in the 1950s, the possibility of there being patches of vegetation on the Moon was still being seriously discussed.

Two problems have always existed for an astronomer observing the Moon – and every other heavenly body for that matter. These are, first, the *physical* restrictions imposed by the quality of any particular telescope and the turbulence of the Earth's atmosphere, and, second, the *intellectual* limitations imposed by dealing with objects far removed from everyday experiences. In the case of the Moon, the physical limits on telescope quality are soon reached. There is so much light coming from the Moon that large telescopes with great light-gathering power are not needed, and the limitations on what can be seen are controlled by the disturbances in the Earth's atmosphere rather than by the quality of the instrument.

The intellectual limitations arose because astronomers fell into the trap of trying to interpret the Moon's surface features too closely in terms of processes they were familiar with on Earth. Their difficulties were compounded by the fact that they were *astronomers*; they had their heads in the clouds (metaphorically), and little knowledge of physical processes operating on Earth, and therefore some of the theories they advanced to account for lunar features were extremely naïve. They would have done better to keep their feet firmly on the ground!

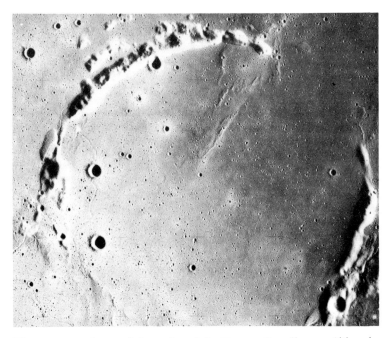

Figure 2.15.   A part of the surface of the Oceanus Procellarum. Although it looks smooth when seen through Earth-based telescopes, this Apollo photograph reveals that every part of the surface is covered with small craters.

Figure 2.16. The splendid crater Tycho in the southern part of the Moon. Note the concentric terraces in the wall of the crater, and the well-developed central peak. There are few craters on the floor of Tycho, indicating that it is itself youthful. (Lunar Orbiter V.)

The subject on which greatest controversy centred was the origin of lunar craters. These are far and away the most conspicuous features of the Moon's surface, and we now know that almost the entire history of the Moon is bound up with the formation of crateriform structures. Figures 2.15 to 2.17 illustrate the range of different types of features included under the heading of 'crateriform' structures. Figure 2.15 shows part of the Oceanus Procellarum, on the near side of the Moon. Hundreds of 'ordinary' fresh-looking craters are visible, the smallest being only a few hundred metres across, and there is one large 'ring' structure about 50 kilometres across with rather ghostly walls. Figure 2.17 shows the opposite end of the size range. The large circular structure is the Orientale Basin on the far side; it is about 700 kilometres across. Most of the maria or 'seas', such as Mare Imbrium, are also large circular basins.

Figure 2.17.   The Orientale Basin on the far side of the Moon. Had this magnificent structure been visible on the near side, speculations on the origins of lunar craters might have followed a quite different course. (Lunar Orbiter IV.)

Clearly, then, any theory for the origin of lunar craters has to be able to encompass an enormous range of structures. Two conflicting schools of thought arose. One, propagated by J. D. Dana in 1846, argued that craters are all of *internal* origin, and were formed by volcanic processes of the kind that form large volcanic calderas on Earth (Figure 2.18). The other, usually considered to have been initiated by G. K. Gilbert in 1893, considered that they had an *external* origin, and were formed by the impact of meteorites. The controversy dragged on through the decades, sometimes flaring up into bitter personal disputes and sometimes nearly flickering out altogether from lack of interest and lack of fresh data.

Both sides took deeply entrenched positions, and were reluctant to admit of any middle ground. All kinds of data were called in

Figure 2.18.   The Earth has few crateriform structures to compare in scale with the lunar ones. This one, the Cerro Galan caldera in north-west Argentina, is about 40 kilometres in diameter, and has a complex volcanic origin. It has a snow-covered central peak. Terrestrial volcanic structures like these were used to argue for the volcanic origin of lunar craters. (LANDSAT *photograph*)

to prop up the arguments of both sides. Analogies were drawn with volcanic and meteorite craters on Earth (often forgetting the unfortunate fact that the different gravitational fields would be bound to produce different structures). Endless statistical analyses were undertaken; the depths, diameters, heights of walls, heights of central peaks, degree of circularity and so on were all measured and plotted against one another *ad nauseam.*

It would take the rest of this book to explore all the various arguments that were raised. Suffice it to say here that no consensus was reached among astronomers prior to the Apollo project. A consensus did begin to evolve during the preparatory stages of the project, which subsequently hardened when the missions were accomplished. The consensus was – perhaps predictably – that both arguments were partly right. Almost every crateriform structure on the Moon is of primary impact origin, but the impacts, and particularly the largest ones, triggered off some impressive volcanic phenomena. While quite small meteorites are capable of producing craters tens of metres across, impacts of bodies the size of asteroids would be required to blast out the major circular structures.

Some other surface features

Although craters are much the best-known features of the Moon's surface, there are others which became the subjects of debate – indeed, they were dragged in as evidence in the arguments over the origins of craters. *Rilles* are the most important of these; they are essentially long, narrow valleys, but three major types were recognized: (a) flat-floored parallel-sided rilles, (b) sinuous rilles and (c) crater-chain rilles.

Figure 2.19.   Hyginus Rille, a magnificent example of a flat-floored, parallel-sided rille. The craters aligned along it have been used as evidence for its origin as a volcanic phenomenon. (Lunar Orbiter III.)

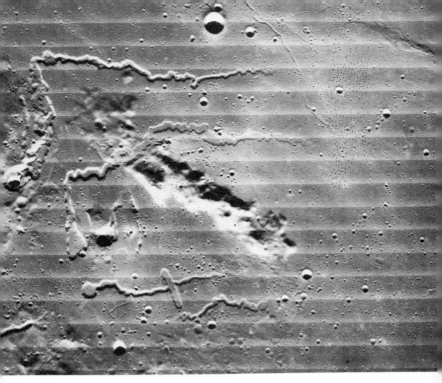

Figure 2.20. Sinuous rilles on the surface of one of the lunar maria. The origin of these rilles is not well understood. Three of the rilles at the centre of this picture definitely seem to originate in craters; one is a compound rille, with a smaller rille nested inside a bigger one. (Lunar Orbiter V.)

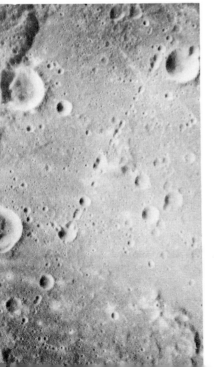

Figure 2.21. A crater chain rille. This line of craters could possibly have been produced by the impact of a stream of ejecta from a distant large impact. (Apollo Metric)

It is surprising what misconceptions grew up among telescopic observers working on rilles. Many of them thought the rilles to be deep, jagged rents in the Moon's surface; the satellite photos show them, by contrast, to be rather shallow, gently moulded features. For the most part, the existence of rilles – especially the parallel-sided and crater-chain types – were used to support the hypothesis that the Moon's surface was shaped by volcanism. It was argued, for example, that the Hyginus Rille (Figure 2.19) was a close analogue of a kind of volcanic rift that is developed on Earth, and that crater-chain rilles are extremely similar to craters developed along a terrestrial volcanic fissure. How on Earth, the argument went, could you explain the formation of a linear chain of craters as the result of random meteorite impacts? In general, the modern view largely bears out the volcanic hypothesis for rilles, but the arguments are by no means closed, particularly in the case of sinuous rilles, some of which look uncomfortably like channels along which liquids have flowed. They may be collapsed lava tunnels, formed originally within moving lava flows. Such tunnels are well known on basaltic volcanoes on Earth, where the longest reach lengths of tens of kilometres. Some crater-chain rilles may also have been produced by *secondary* impacts of streams of ejecta thrown up when neighbouring craters were blasted out by primary impacts (Figure 2.21).

Volcanoes

It was obvious even to the most hard-bitten proponents of the volcanic theory of crater origin that there are no volcanoes of terrestrial type on the Moon; no sweeping cones of the Mount Fuji type, no great Hawaiian shields, nothing. This was puzzling, particularly because of the fact that there was plenty of observational evidence to suggest that the great grey 'seas' were covered with enormous expanses of basalt lavas. Where had all these huge volumes of lava come from?

The nearest lunar approach to terrestrial volcanoes are some rather odd, blister-like swellings or *domes* which rise above the

Figure 2.22.   These broad, swelling domes are located near the crater
Gruithuisen on the western shores of the Mare Imbrium. They are thought
to be of volcanic origin, a hypothesis which is supported to some
extent by the presence of summit craters on them. (Lunar Orbiter V.)

surfaces of the maria. Some of them are large, reaching up to 20 kilometres across, and some have summit craters. It is now generally thought that they are true volcanic phenomena, but are extrusions of viscous acid volcanic rocks rather than volcanoes in the conventional sense. The fluid basaltic lavas which flood the 'seas' are thought to have welled up from rather inconspicuous fissures around the edges of the 'seas'.

During the many decades of telescopic study of the Moon, there were hundreds of 'observations' of flashes of light, of bright glows, of red clouds and other mysterious, short-lived events. These were, and are, dignified with the title 'transient lunar phenomena', or TLPs for short. For many years it was claimed that many of these were observations of volcanic eruptions taking place on the Moon, and that others were the result of meteorite impacts. The volcanic hypothesis received considerable support in 1958 when the Russian astronomer, Nikolai Kozirev, claimed to have detected an emission of a reddish cloud of hot gas lasting half an hour from a small crater near the summit of the mountains at the centre of the crater Alphonsus. He also produced a spectrogram showing that there had been a temperature rise of 2,000 degrees, and that the gas contained carbon in some form.

Although widely accepted at the time, no one has been able to confirm Kozirev's work, and the idea that there might be active volcanism on the Moon has been almost entirely discredited by the discoveries of the Apollo landings. Most TLPs, in fact, appear to be in the eyes of the observer.

Rays

Seen through a telescope, or even a humble pair of binoculars, the full Moon is a spectacular sight. The most arresting features are the whitish streaks or 'rays' which radiate out from a few prominent young craters such as Tycho. The rays cannot usually be traced back right into the crater centres, but rather to an 'area of confusion' defined by the crater walls. In most cases, they are brighter than their surroundings, but in a few cases they are darker. The differences in brightness have been attributed both to differences in composition of the material making up the rays and to differences in its grain size. The rays are remarkable for their great length – some of those from Tycho extend for over 2,000 kilometres.

Not surprisingly, the rays have been the subject of much dispute, and rival explanations were offered by the volcanic and meteorite impact schools. The impactists argued that the rays

Figure 2.23. A striking example of a rayed crater. This one is about 10 kilometres in diameter, and is located on the far side of the Moon. Note the pale colour of the streaks, and the fact that, while they radiate out over long distances, they are patchy and discontinuous. (Apollo Hasselblad)

were the ejecta produced by an impact event, or possibly fracture systems produced by the impact; the volcanists argued that the rays were either the ejecta caused by a catastrophic volcanic eruption, or gaseous condensates escaping from fractures produced by the eruption.

Both sides agreed that the rays were probably composed of particles scattered in a thin film over the lunar surface, but the impactists thought that the particles were probably glass beads produced by cooling of molten material ejected at high velocities, whereas the volcanists thought that the particles were ash fragments analogous to terrestrial pumice.

Apart from the fundamental nature of the rays themselves, there were two other factors that required explanation. First, there are significant variations in the brightness of rays associated with different craters; and secondly, there are obviously no rays associated with old craters – they are confined to the youngest visible structures. The variation in brightness of the rays could be easily explained on either hypothesis; material ejected from different impact sites was of different composition, or the material erupted from different volcanic centres was of different composition. The absence of rays from old craters was more difficult to explain, although some of the volcanists tried to brush it away by saying that the older craters simply did not erupt any ash. The correct explanation, that the rays are thin streaks of impact ejecta, and that the ray material becomes progressively darker with time as a result of exposure to solar radiation, was not fully appreciated until the Apollo programme was under way. The fact that ray systems become progressively less visible with age formed an important part of the first geological studies of the Moon (to be discussed in Chapter 3).

The Moon's atmosphere

One does not need to be an astronomer to deduce that the Moon has an exceedingly thin atmosphere. Night after night it shines unvaryingly down on us, its surface never masked by the spiralling cyclones of cloud that soften views of the Earth seen from space, nor by the yellowish veils of dust that often obscure the surface of Mars. With a telescope, further limits on the extent of the lunar atmosphere are obvious. If a star is observed close to the Moon, such that the Moon will move in front of it and *occult* it, the star shines steadily and fixedly (apart from effects created by the Earth's own atmosphere, of course) until it winks out sharply behind the Moon. Had even a thin atmosphere been present, the light from the star would have been refracted by it, and this would have been clearly discernible, not only in the 'twinkling' of the star but also in the precise time of its occultation.

These observations are only to be expected from theoretical considerations. As outlined in Chapter 1, given the size, mass and surface temperature of a planet, it is possible to calculate its escape velocity and the density of gases which it could retain. The mass of the Moon is quite small, and the surface temperature is very low. It follows that only dense gases could be retained, such as carbon dioxide and sulphur dioxide. Some telescopic attempts were made to detect these gases, but the results showed that if they were, indeed, present, they must be there in vanishingly small amounts. It was shown, in fact, that the density of the lunar atmosphere could not be more than one ten thousandth of the Earth's sea-level atmospheric density.

Life on the Moon

All the telescopic studies of the Moon confirmed that it is thoroughly inhospitable: a bleak, airless world, quite unlike the Earth, and a most unpromising place to look for life. This did not forestall serious discussion about the existence of life on the Moon. Looking back, it seems scarcely credible that its existence there could ever have been seriously considered, especially since the telescopic observations of the surface could be backed up with precise estimates of surface conditions. In astronomical terms, the Moon is extremely close to the Earth, and an easy object to observe. There has none the less been a consistent belief in some quarters in the existence of life on the Moon, and a strong desire to find – one is tempted to say, to manufacture – evidence for this belief. A similar but much more serious situation grew up with Mars.

Perhaps these desires to find life beyond the Earth are an expression of a basic weakness or sense of insecurity in mankind – he is reluctant to believe that he is alone in the solar system. Whatever the reason, it has led both scientists and the lay public into accepting at face value some pretty tall stories.

A splendid example is the so-called 'lunar hoax' of 1835. At that time, a famous astronomer named John Herschel (the son of the great William Herschel) sailed to the Cape of Good Hope to make some new observations on southern hemisphere stars. While in Africa, he was more or less completely out of touch with the world. An unscrupulous New York journalist named Richard Locke seized on Herschel's absence as a golden opportunity to pull the scientific community's leg. He wrote an article in the New York *Sun*, in which he claimed that Herschel had devised a new kind of telescope with enormous powers of magnification which enabled him to observe some remarkable forms of life on the Moon. He provided a highly detailed description of the fabulous telescope, and furnished many lurid details of what was supposed to be on the Moon: amethyst mountains, sapphire hills and rock columns of green basalt constituting a landscape which was populated by flying unicorns, ape men 'who averaged four feet in height and were covered, except on the face, with short and glossy copper-coloured hair and had wings composed of a thin membrane', and even more exotic life-forms, such as 'the strange amphibious creature of spherical form, which rolled with great velocity across the pebbly shore'.

This all seems absurd enough to us now, but at the time it was accepted readily by both the public and scientists alike. One New York newspaper commented, 'These new discoveries are both probable and plausible', while another considered that the observations had 'created a new era in astronomy and science generally'. It was a full year before Herschel got to hear of the hoax perpetrated in his name, and published a refutation. That, of course, was over a century and a half ago, and both scientists and public grew slightly more sceptical in the decades that followed.

Only slightly, though. In the 1920s, the distinguished American astronomer, Professor W. H. Pickering, was observing the crater Eratosthenes. He noticed a number of small dark patches which seemed to vary regularly in position with each lunar 'day'. Although he had previously argued that vegetation existed on the Moon, he proposed in a paper published in 1924 that these dark patches were swarms of *insects*, and furthermore, that the movement of the patches was a result of the wholesale migration of the insect swarms in response to changing conditions during the lunar day. To reinforce his argument, he pointed out that, had an astronomer on the Moon in the nineteenth century pointed his telescope at the great plains of America, he would have seen shifting dark patches caused by the migration of buffalo herds!

Perhaps feeling that this startling suggestion might not go down too well, Pickering wrote:

> While this suggestion of a round of lunar life may seem a little fanciful, yet it is based strictly on the analogy of the migration of the fur-bearing seals of the Pribiloff Islands . . . The distance involved is about twenty miles, and is completed in twelve days. This involves an average speed of six feet per minute, which, as we have seen, implies small animals.

Pickering was either *extremely* gullible, or else had a well-developed sense of humour and was gently poking fun at some of his colleagues. We shall never know.

More recently, several astronomers observed – or thought they observed – radial dark bands in some craters, most notably Aristarchus. In 1951, Patrick Moore, a household name in Britain and a respected student of the Moon, proposed that the dark bands were created by primitive forms of vegetation, clinging precariously to life in the favourable microclimates produced around deep radial cracks in the crater floor. Wafts of vapour exhaled from these cracks would suffice to sustain the life of the vegetation. Moore was aware that, whatever life-form might exist, it would have to be quite unlike anything on Earth, and extremely primitive. None the less he was prepared to believe that life existed in the face of a vast weight of evidence to the contrary. Primitive forms of vegetation are a far cry from 'amphibious creatures of spherical form', but they demonstrate that belief in life on the Moon was still alive and flourishing during the decade before the Apollo 11 landing.

A brief lunar re-capitulation

From the summer day in 1609 when Thomas Harriot turned the first primitive telescope on the Moon until President Kennedy's initiation of the Apollo project in 1961, many tens of thousands of hours were spent observing the Moon, by astronomers in all parts of the world using instruments ranging from small home-made affairs to the giant 200-inch monster at Mount Palomar. But how did these studies contribute to our knowledge of the Moon?

The size, shape and density were all well known – the telescope served mainly to refine these. From knowledge of the density, inferences could be drawn about the composition. Radar studies provided further data on the composition and structure of the surface layers. It scarcely needs a telescope to demonstrate the absence of an atmosphere. In fact, telescopic work provided a mass of detail about the *geography* of the Moon, and something about surface conditions, but little else. It could say nothing about the internal structure of the Moon, or its internal temperature, its age, chemical composition or magnetic field.

Thus, despite the huge amounts of time invested in telescopic work, our knowledge of the Moon prior to the Apollo project was sketchy in many important areas. Constructing detailed maps of surface features does not really help us to understand their nature – the controversy over the origin of lunar craters was still deadlocked prior to Apollo, and there were still lingering beliefs that primitive forms of life might exist. It would be useful to bear this perspective in mind in considering the Apollo results. It would be easy otherwise to lose sight of the magnitude of the achievement in the wealth of detail.

3. Project Apollo

The Apollo project culminated in what is widely regarded as the greatest achievement of technology and science in history. Unfortunately, it was not conceived from purely technological or scientific motives. When Neil Armstrong made his 'great leap for mankind' on 20 July 1969, it was the ugly imperatives of the ideological struggle between West and East rather than the quest for knowledge that enabled him to make that leap. In this respect, Man's exploration of the Moon was no different from his earlier great voyages of discovery, most of which, including those of Columbus, were motivated more by the lure of quick profits than any desire to expand the sum of human knowledge.

Reams have been written about the political and economic aspects of the space race. Since there is so much to say about the scientific achievements of Apollo, these controversial topics will be ignored here, save for a brief historical review.

In 1957, the USSR launched Sputnik 1, the first artificial satellite. In 1958, the USA launched Explorer 1, their first satellite, and much smaller than Sputnik 1. By 1959, the USSR had succeeded in sending a spacecraft round the far side of the Moon, and in obtaining some fuzzy pictures of it; by 1961 they had sent Yuri Gagarin into space to achieve a single orbit in Vostok 1. The best response the Americans could make to this was the sub-orbital hop of Alan Shepherd, in a Mercury capsule in the same year. Clearly, the Americans had fallen far behind the USSR and were losing the propaganda battle hands down.

It is all the more amazing that they were able to work their way up from a position so far behind, and, in a period of less than ten years, make enormous strides of progress. This was achieved only by the dedicated efforts of many thousands of talented individuals working together. As President Kennedy said in the speech to Congress which initiated the Apollo project: '. . . in a very real sense, it will not be one man going to the Moon . . . it will be an entire nation. For all of us must work to put him there.'

Much of the criticism of the Apollo project has been concerned with its enormous expense. The huge sums of money could better have been spent on feeding the underdeveloped countries, the protesters cry. Some facts may be useful for those who wish to debate the issue: in the years 1959–72, which covered the Apollo project and its lead up, the total NASA budget was $46·8 billion. The total federal budget for the year 1972 *alone* was $266 billion, of which $72 billion was spent on health, welfare and education. In 1976, the American people spent nearly one billion dollars on chewing gum, and $20 billion on alcohol.

Endless discussion has also centred on the decision to send men to the Moon, rather than to use automated spacecraft. Space does not allow us to rehearse these arguments here; the fact is that manned missions were sent, and it is these that will be described. It will become apparent in the later chapters on Mars and Venus how much can be achieved by unmanned spacecraft, and the reader may then form his own view. Almost the whole of the present chapter will be concerned with the Apollo project, and little will be said about the Russian programme of lunar exploration. This is not because of any political bias, but the simple fact is that the Russians have obtained much less data and have not published all which they have obtained. By contrast, there is such a flood of data from the Apollo project that it is difficult to cope with it all.

The preparatory missions

Prior to Apollo 11, there were eighteen preparatory unmanned lunar missions, and three series of manned missions: Mercury, Gemini and the early Apollos. The manned missions were progressively more sophisticated tests of spacecraft designs and systems;

the last of them, Apollo 10, was a complete dress rehearsal, with every stage except the final lunar landing being carried out.

The unmanned missions were for the purposes of reconnaissance, to ascertain what the first man to step out on the surface of the Moon would find there, or, indeed, if he could even step on to the surface at all, rather than disappearing into deep drifts of dust as some scientists had gloomily predicted. The unmanned missions in themselves represented an enormous advance over pre-existing knowledge of the Moon.

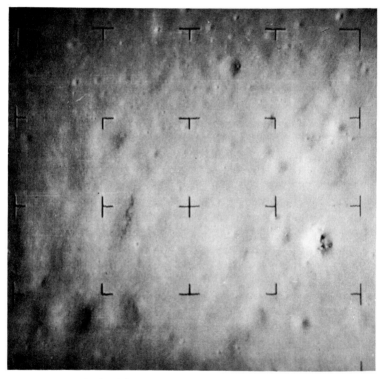

Figure 3.1.   The final frame taken by the Ranger 7 spacecraft before it crashed on to the Moon on 31 July 1964. Pictures like these provided final confirmation that the lunar surface presented no serious obstacles to landing spacecraft. The smallest craterlets are less than a metre across.

Ranger

The Ranger series was designed to provide the first close-up pictures of the Moon's surface, to see if it was smooth enough to attempt a landing. The Ranger craft, however, were bedevilled by many technical failures, and it was not until the seventh mission in the series in 1964 that pictures were successfully obtained. These showed progressively more detailed views of the Mare Cognitum as the spacecraft approached; the last frame transmitted before the spacecraft smashed into the surface had a resolution of about half a metre. Two other missions followed, both of them returned high-quality pictures and both successfully crashed – or 'hard-landed', as the Americans say – at their designated sites.

The Ranger pictures showed features which had never before been visible. The best terrestrial telescopes could not resolve features less than 500 metres across; the Ranger pictures showed details less than a metre across. They revealed few major surprises, however. They showed that the surfaces of the maria were pockmarked with myriads of tiny craters and craterlets, just as radar studies had suggested, and that the mare surfaces were certainly not so rugged as to prove a major problem for landing missions.

Lunar Orbiter

As their name indicates, the Lunar Orbiter missions were designed to orbit the Moon for prolonged periods, rather than smashing straight into it as the Ranger craft had done. Their objectives were to obtain high-quality pictures of as much of the Moon as possible, so that it could be precisely mapped by both topographers and geologists, and so that landing sites for the Apollo missions could be chosen. Additionally, they were used to establish the intensity of radiation in the vicinity of the Moon, to make gravitational studies and to monitor the number of micrometeorites likely to be encountered during the Apollo missions.

(There was still some worry at that time that high-velocity specks of cosmic dust could represent a hazard to space travellers by drilling tiny holes through their space suits and through them.)

The Orbiter missions were amazingly successful, and contributed an enormous amount of new data on the Moon – so much so that, a decade later, much interpretive work still remains to be done. Five Orbiter missions were flown. By the end of the series, 99 per cent of the Moon had been photographed at a resolution of 60 metres or better, and smaller areas had been covered at a resolution of about 1 metre.

The cameras on board the spacecraft used 'ordinary' 70-mm Kodak film designed for aerial photograph work, which was developed on board. Instead of being dunked in baths of chemicals, as in a home darkroom, the exposed film was pressed against a pad soaked in the appropriate developing and fixing solutions – essentially the same kind of process used in black and white 'Polaroid' cameras. The film negative was dried, and then electronically scanned, and the signals transmitted back to Earth. The scanner could only process strips 2·54 millimetres wide by 60 millimetres long, so each strip was transmitted separately, and the final photographs had to be reconstructed from dozens of individual strips. This makes the final product look rather disjointed, but it was done in the interests of higher resolution. Each strip took twenty-two seconds to scan, and each has 17,000 scan lines. Compare that with the 625-line scan of an ordinary TV set!

Despite their strip structure, the quality of the Orbiter pictures was stunningly high, as many of the illustrations used in this chapter testify. For the first time, scientists had a clear view of the lunar surface, unsullied by the murk of the Earth's atmosphere. Their task of selecting landing sites was made relatively simple, and, for the later missions, they were able to take geological factors revealed by the pictures as well as safety considerations into account.

Figure 3.2   An example of the magnificent photographs obtained by the Lunar Orbiter series of satellites. The large crater is Copernicus. A comparison of this picture with Figure 2.11 (page 53) demonstrates convincingly the limitations of terrestrial lunar photography. (Lunar Orbiter IV.)

The micrometeorite studies were encouraging. They showed that, during all five missions, involving a total of many weeks in lunar orbit, only twenty-two impacts were registered on the spacecraft, and these were of minute particles. Hence, no hazard existed for the Apollo craft that would have to spend only a matter of days in orbit around the Moon.

On a purely scientific level, the gravitational studies yielded some important results. The studies were made by observing

Figure 3.3.   Variations in the Moon's gravitational field as determined by Lunar Orbiter tracking data. The obvious circular 'highs' correspond to the circular mare basins, as examination of Figure 2.6 (page 45) will confirm.

changes in velocity of the spacecraft, which could be measured in terms of changing frequencies of radio signals transmitted from them – an application of the effect known as Doppler shift. When the spacecraft passed over regions of the Moon where the acceleration caused by gravity was higher, they speeded up slightly; when they passed over regions of lower gravity, they slowed down. P. M. Muller and W. L. Sjogren at the Jet Propulsion Laboratory in Pasadena, California were able to show that the maria are almost all areas of high gravity (Figure 3.3).

These *positive gravity anomalies* indicate that local excesses of mass are present: large lumps of material more dense than their surroundings. This discovery was immensely interesting to geologists, because it supported the long-held idea that the maria were basins filled with relatively dense, basalt lavas, and, more important, that the Moon probably had a thick, rigid crust.

Both positive and negative gravity anomalies exist on Earth, but they tend to be relatively short-lived affairs (by geological standards) because of the phenomenon known as *isostasy* – dense parts of the crust tend to sink, and light parts to rise, until inequalities in mass are smoothed out. A good example is the area of the Baltic Shield. This was depressed by the weight of ice sheets during the Ice Age, forming a marked negative gravity anomaly. When the ice sheets melted some 10,000 years ago, the land rose again, and is still rising. Soon the negative gravity anomaly will disappear. On the Moon, no such compensation process appears to take place.

Surveyor

The Surveyor series of spacecraft were small-scale flying test beds, designed to investigate the engineering problems of landing on the lunar surface. Each weighed only 100 kilograms, and consisted of a spidery triangular framework of aluminium tubing, to which were attached retro-rockets, solar panels, a TV camera and so on. Seven of these ungainly craft were launched, and six successfully soft-landed.

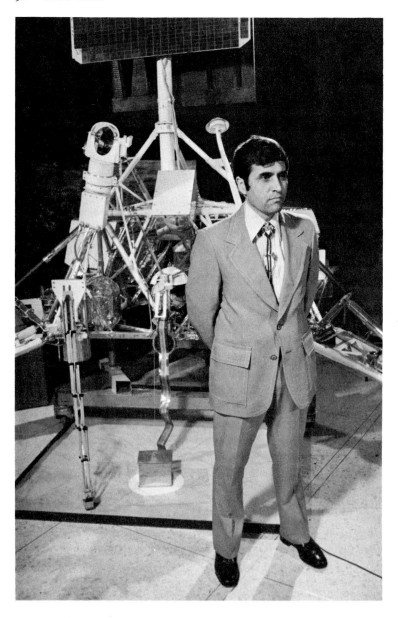

Figure 3.4. Astronaut Harrison 'Jack' Schmitt and one of the Surveyor series of soft-landing spacecraft. The spacecraft camera is mounted in the white housing on the left, fitted with a periscope mirror. The box on the floor contains the alpha-backscatter instrument, and to the left of that is the surface scraper. Schmitt flew on the Apollo 17 mission. He is now Senator for the state of New Mexico.

Surveyor 1 established a number of 'firsts': the first close-up pictures of the lunar surface *from* the surface; the first soil mechanics experiments; the first chemical analysis; and so forth.* The TV camera on Surveyor 1 showed a bleak landscape – a dreary, flat surface pitted with small craters and with a few small boulders lying around. More important, the camera was able to look at one of the footpads of the spacecraft, and to show that, while it had sunk in a little, the surface was clearly strong enough to support the weight of the much heavier Apollo lunar module.

This important result was confirmed at the other Surveyor landing sites. In one case, the descent engines of the craft were fired briefly after landing so that the craft made a short hop. The television pictures showed that the footpad had left a sharp footprint, indicating that the surface material was very coherent – the footprint was like that a man would leave when walking on firm, damp, beach sand. Two of the craft also carried an extraordinary surface scraper which was basically a scoop at the end of a long trellis-like affair. This could be extended from the spacecraft by remote control and used to scrape a shallow trench in the surface, an experiment again designed to investigate lunar soil mechanics and to confirm that there would be no unexpected hazards to a manned landing – no hidden cavities for an astronaut to stumble into.

Scientifically, by far the most important achievements of the Surveyor series came from the *alpha backscatter experiments* carried on the last three flights. These instruments worked by irradiating a small patch of the lunar surface with alpha particles from a

* The first ever soft landing was made by the Russian Luna 9 six months earlier. It also obtained some pictures, but worked for only three days.

*Figure 3.5.   The lunar
landscape as seen from the
Surveyor VII spacecraft,
looking out over a wilderness
of small rocks and craters 29
kilometres north of the crater
Tycho. The small craters
and blocks in the foreground
may be secondary features,
produced by the impact
which excavated a much
larger crater in the region.*

radioactive source and analysing the radiation bounced back; the *energies* of the backscattered particles were characteristic of the elements present in the surface material while the *numbers* of particles at different energy levels gave an idea of the relative abundances of the different elements.

The alpha backscatter experiments therefore made it possible for the first time to analyse directly – albeit crudely – the surface material of an object other than the Earth. The results themselves were not particularly surprising: they confirmed a basaltic composition for the rocks of the maria. The data for the highland regions of the Moon were rather different, but discussion of that is best deferred until later (pages 134–6).

Lunar geological mapping

Perhaps the single greatest advance during the period leading up to the Apollo missions was made using telescopic data, not spacecraft. In 1962, Eugene Shoemaker and Robert Hackmann of the US Geological Survey published a twelve-page paper of breathtaking scope and daring, entitled simply 'Stratigraphic Basis for a Lunar Time Scale'. This was an immensely ambitious undertaking, especially since their initial work was based on a single photograph of the crater Copernicus, taken with the Mount Wilson 100-inch telescope (Figure 2.11, page 53). However, the time was ripe, and, as Shoemaker and Hackmann boldly stated in the first line of their paper: 'The impending exploration of the Moon and geological mapping of its surface raise the need for an objective lunar time scale.'

Shoemaker and Hackmann's work marked a watershed in the history of lunar science, greater, perhaps, than even the Apollo landings. Overnight, the publication of the paper removed the study of the Moon from the hands of the astronomers, who had held it for centuries, and placed it firmly in the hands of geologists.

Briefly, Shoemaker and Hackmann showed that rocks of the Moon could be assigned to different age groups by determining which rocks are lying on top of which. This is the basic concept of terrestrial stratigraphy, and led to the recognition on Earth of the geological systems which will be familiar to many readers – Ordovician, Silurian, Jurassic and so on. The lunar systems – Imbrian, Eratosthenian, Copernican and so forth – will be less familiar, but they were established on the same principles.

After the publication of Shoemaker and Hackmann's pioneering work, geological mapping of the whole of the near side of the Moon was undertaken by the US Geological Survey. Their ultimate aim was the production of forty-four quadrangle sheets at a scale of one to one million, though much more detailed maps were prepared for the Apollo landing sites and back-up sites.

Associated with the mapping work were other extensive studies of the relative ages of different parts of the lunar surface. Consider what happens in a fairground shooting gallery. Behind the targets there is usually a sheet of metal against which the spent bullets clang. Now, when such a sheet is new, it will be shiny and virginal. After one night's shooting, it will be pitted with small dents scattered all over its surface. (If the rifles are good, and the customers not giggling girls, the distribution may not be random.) After a second night, the surface will be more pitted; on the third, even more so. Eventually, a point will be reached at which the surface is saturated with bullet marks; a new mark can only be made by obliterating an old one.

A somewhat similar situation exists on the Moon. If a surface is exposed for long enough to meteorite bombardment, it will eventually become so heavily cratered that it is saturated. *Before* it becomes saturated, though, one can get an idea of how old a surface is by counting the density of craters; one with only a few craters must be young, a heavily cratered one must be older. This is the basis for *crater density* studies on the Moon, but there are two important complications. First, not all craters are the same size; and secondly, one cannot assume that the rate of cratering has always been constant. The first problem is easily dealt with: one merely counts the numbers of craters within particular size ranges per unit area, and plots a graph for the surface concerned, as in Figure 3.6. The second problem is much more difficult, and is still being grappled with by a number of scientists.

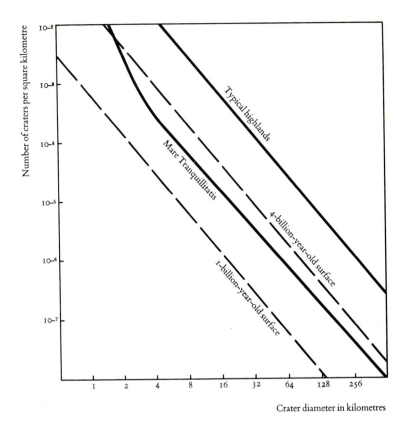

Figure 3.6. There is a logarithmic relationship between the size and frequency of lunar craters; clearly there are many more small craters than large ones. The measured distributions for highland and mare areas are shown in solid lines; note that the highlands are 'saturated' with craters. The dotted lines show the calculated distributions for surfaces 4 billion and 1 billion years old. (After W. K. Hartmann)

By making simple size/frequency distribution studies, then, it is possible to establish quite subtle differences in ages between different parts of the Moon's surface. The most obvious difference is between the light-coloured highland areas, which are so heavily cratered that they have reached 'saturation', and the dark-coloured maria, which have hardly any large craters and therefore must be much younger. These studies of cratering on the Moon are of first importance because, on the Moon, scientists have been able to sample the surfaces, *measure* their ages directly and thus 'calibrate' the technique of crater counting. For example, it is a reasonable first assumption that an area on Mars with the same crater frequency distribution as one of the lunar maria is of a similar age.

Project Apollo

According to one of the Apollo astronauts, the question which he was most often asked on his return to Earth was, as he called it, the WWILOTM question, or, 'What Was It Like On The Moon?' This, apart from illustrating the delight everyone associated with the Apollo project had for coining ugly acronyms, is indicative of the rather limited impact that the Apollo missions had on the public – the astronauts were rarely asked about the scientific objectives of their missions.

If readers are interested in the details of spaceflight – and who could fail to be – they will find many sources of information. In terms of pure gripping readability, by far the best is *Moonwreck* by H. S. F. Cooper, a minute-by-minute account of the ill-fated Apollo 13 mission. As usual, it is when things go wrong that 'human interest' reaches a maximum. No one has written a book about the flawless Apollo 17 mission!

The scientific discoveries of Apollo are, in their way, just as fascinating as the human adventures. The major problem in trying to describe them is that there is so much to cover. People tend to think that the only scientific work done on the Moon was that actually carried out by the astronauts while they were there. It is often forgotten that, while the lunar modules were on the Moon, the command service modules remained in orbit, with the instruments on board continually scanning the lunar surface and providing remote sensing data over huge areas. Furthermore, apart from collecting samples and making their own observations, the astronauts set up on the Moon instrument packages which went on working long after they had left. One of them continued to send data down to Earth for more than *ten years*, and was still going strong when the monitoring equipment on Earth had to be closed down for lack of funds.

The orbital experiments

Of the many studies carried out in lunar orbit, the photographic work was the most straightforward, and has the most immediate impact. The quality of many of the pictures is outstanding, far better than even the superb Lunar Orbiter views. This is, of course, because the Apollo photographs were taken by hand, using the very best available cameras (Hasselblad) and ordinary film which could be returned to Earth for processing.

Some of these photographs have become among the best known in history. Perhaps the most profoundly moving of all are those which show the blue Earth, wreathed in white clouds, rising above the drab, cratered wastelands of the Moon. It has been argued that these photographs were a major contributing factor in the great change that took place in human outlook in the 1970s, when people suddenly became aware of how small the Earth is, and how easy it would be to upset the environmental balance that makes 'Spaceship Earth' inhabitable.

The enormous aesthetic and sociological impact of the pictures, however, was an unexpected by-product of their original purpose, which was to make more detailed studies of the lunar surface than could be achieved with the Orbiter pictures. The astronauts were also able to take pictures of selected features that caught their eye when the angle of illumination was favourable.

The fact that the Moon's 'atmosphere' is equivalent to a hard vacuum on Earth meant that scientists could attempt some kinds of remote sensing experiments that would be impossible on Earth. Two of these were connected with finding out more about what lunar rocks are made of, and how their compositions vary from place to place. These studies were of considerable importance because, although the successful landing missions returned a

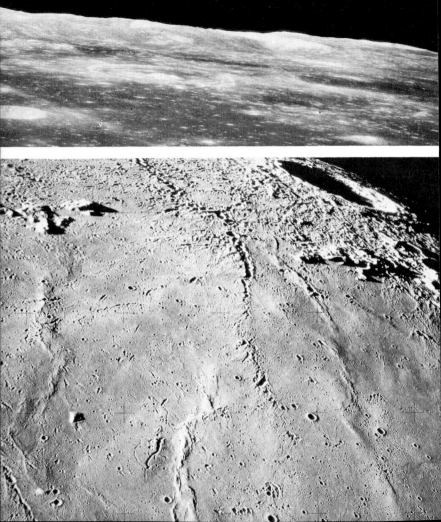

Figure 3.7. (opposite)   *The Earth seen from the Moon.*
(Apollo 11, Hasselblad)

Figure 3.8. (opposite below)   *A striking oblique view of Copernicus*
(top right). This one was taken by an Apollo 17 astronaut with a hand-held
Hasselblad camera. The area shown lies to the north-east of Copernicus.
Compare this view with Figures 3.2 (page 87) and 2.11 (page 53).

Figure 3.9.   The Apollo 17 command and service modules, seen from
the lunar module in lunar orbit. The open compartment in the upper part
of the spacecraft is the scientific instruments module (SIM) bay, used in
carrying out remote sensing experiments. (Apollo Hasselblad)

flood of data of a much more tangible form – including lumps of actual surface material – these data necessarily related only to small parts of the Moon. So it was highly desirable to use remote sensing techniques to try to extend the observations made at the landing sites to the whole of the Moon.

One of the most useful techniques used in terrestrial geochemical laboratories is known as *X-ray fluorescence*, or XRF. Specially prepared rock samples are exposed to X-rays in a hard vacuum. This radiation excites the atoms present in the rocks, which then re-radiate X-rays at different wavelengths, the wavelengths being characteristic of the individual elements present, and the intensity proportional to their abundance.

In lunar orbit, the astronauts could not simply point an X-ray source at the surface – far too much energy would be required to carry any such instrument on board. It was also totally unnecessary, because the Moon is permanently bathed in a flood of powerful X-rays – from the Sun. The Earth also receives its share of solar X-rays, but, fortunately for us, these are blotted out by our thick atmospheric blanket. Thus the instrument carried on Apollos 15 and 16 was merely an X-ray detector capable of measuring the intensity of X-rays emitted by lunar surface rocks in response to the glare of the solar X-rays.

This technique enabled the Apollo scientists to map out the relative abundances of aluminium, magnesium and silicon on the Moon. Figure 3.10 shows one of the profiles obtained; it reveals the strong chemical contrast between the smooth mare areas and the paler-coloured cratered highlands, a contrast which is fundamental to the geology of the Moon.

The XRF studies were concerned with X-rays induced by solar radiation. The *gamma ray spectrometer* was simpler in concept, and was designed to measure the *natural* radioactivity of the rocks of the lunar surface by measuring the intensity of gamma rays reaching the spacecraft. This enabled the scientists to establish roughly how abundant such radioactive elements as uranium, thorium and potassium are in the rocks of the Moon's crust. On Earth, the radioactive elements, which are always present in tiny amounts, tend to be concentrated by geological process in

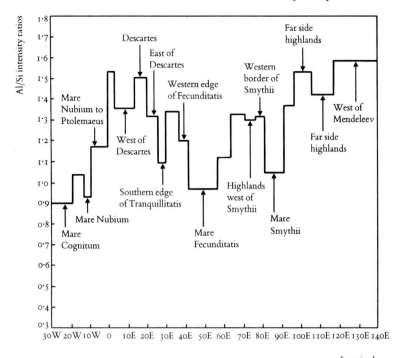

Figure 3.10. The results of one of the Apollo orbital X-ray studies of the lunar surface. Differences in the compositions of the surface rocks are manifested in their aluminium to silicon ratio. The dark-coloured 'seas' all have conspicuously low ratios compared with the pale highland areas.

specific kinds of rocks, particularly rocks characteristic of the continents, such as granites. The lunar results showed that the radioactive elements are much less abundant on the Moon than on the Earth, and that those which are present are not uniformly distributed. These anomalies remain to be explained.

One of the things taken for granted here on Earth is the fact that the Earth has a powerful magnetic field, so powerful that a toy bar magnet suspended on a piece of string will immediately orientate itself north–south. This is something we all grew up with, and probably rarely think about. Even more rarely – if ever – do we think about whether a magnet would do the same on the Moon, or the other planets. This is rather a pity, because magnetic fields are interesting things which can reveal a great deal about the interior workings of planets.

The Moon's magnetic field was investigated initially by Explorer satellites launched from Earth. Much more refined data were obtained by small sub-satellites launched into lunar orbit from the Apollos 15 and 16 spacecraft. These studies showed that the *present* lunar magnetic field is *dipolar*, which means that north and south magnetic poles do exist on the Moon, and so, in principle, a compass should work there. The Moon's magnetic field, however, is exceptionally weak, being only about one twelve thousandth of that of the Earth, which means that it would require a very sensitive compass indeed to seek out the north pole. Most terrestrial ones would be useless. Much more was learned about the *past* magnetic history of the Moon from the returned samples. Most importantly, these showed that the Moon's magnetic field was far more intense in ancient times than it is now.

The Apollo landings

When all the multitude of engineers, communications experts, doctors and flight crews were fully satisfied that the Apollo landings could be safely achieved, a further major question remained to be answered. Where should the spacecraft land? Each mission would cost millions of dollars, so it was absolutely essential that the maximum scientific return be obtained from each of them. Although it may not seem so when one sees it reflected in a quiet lake on a warm, velvety night, the Moon is a big place. Its surface area is comparable to that of the Pacific Ocean, which covers almost a complete hemisphere of the Earth.

The scale of the task facing the site selection team was enormous. Thousands of potential sites had to be screened and sifted, and long debates took place between the proponents of various candidate sites. With Apollo 11, the issues were relatively straightforward – the overwhelming priority was to get the astronauts down in one piece. From the point of view of saving fuel in the spacecraft, the site had to be as near the lunar equator as possible; and for crew safety, the site had to be as smooth and as far from any mountainous terrain as possible. The need to keep in touch with Earth, of course, meant that the site had to be on the near side of the Moon. The chosen site on the Sea of Tranquillity filled all these requirements ideally, but several other sites on the mare surfaces would have done just as well.

After the success of Apollo 11, the later missions could be planned with less emphasis on the safety of the landing sites, and more on the scientific pay-off that could be achieved. Decreasing the number of physical restraints on the sites, of course, made more and more options open to the site selection teams, and their task of choosing more and more difficult. Some of the factors they had to consider were practical: could the astronauts walk, or drive in the Lunar Rovers, to sufficient places of interest? Others were technical: would the sites be far enough from each other to provide an adequate network for those experiments that required that kind of data? And would the sites provide representative samples of all the major lunar rock suites? (It would be pointless to land the astronauts slap in the middle of an enormous lava flow, where they could see nothing but lava from horizon to horizon.)

In making their decisions, all that the site selection teams had to go on initially were the topographical and geological maps constructed from Lunar Orbiter pictures. Later, the first landings produced an element of 'ground truth' which could be used to revise and clarify the photographic interpretations. Broadly speaking, two schools of thought grew up among those involved in the site selection business. One argued that, with so many potential sites to visit, each was bound to provide a mass of new data, so it didn't really matter too much which were selected. The other went to the opposite extreme, and required minute inspection of the pros and cons of each site.

Figure 3.11 shows the sites eventually chosen. Apollo 12 was located in terrain like that visited by Apollo 11, but close to the Surveyor 3 site, since the mission plan called for the retrieval of parts of the Surveyor craft which had been exposed to the lunar environment for over two years. (It is a tribute to the accuracy of

Figure 3.11.   The Apollo landing sites. To conserve spacecraft fuel, preference was given to sites near the lunar equator. Only Apollos 15 and 17 were sited well away from the equator.

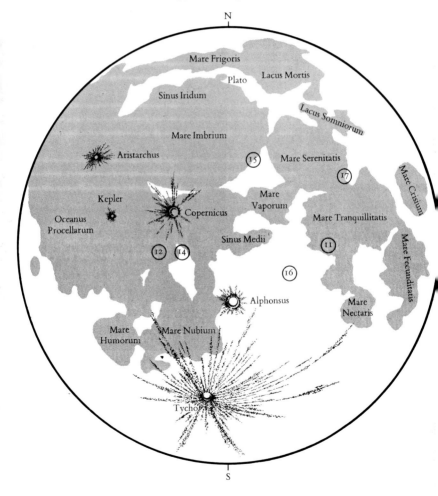

the Apollo navigation systems that Apollo 12 was ultimately landed within walking distance of the Surveyor craft.)

Apollo 13 was aborted – a fact which must have added weight to the superstitions about 'unlucky 13'. The Apollo 14 site was selected in order to study material known as the Fra Mauro

Figure 3.12.   An impressive demonstration of the technical skills evolved during the Apollo missions: the Apollo 12 lunar module stands on the skyline within walking distance of the Surveyor 3 spacecraft. From lunar orbit, Surveyor would be an invisible speck, lost in the vast expanses of the Oceanus Procellarum. (Apollo Hasselblad)

formation, which, it was believed, was ejected during the enormous impact event which created the Mare Imbrium. Apollo 15 was sent to a spectacular site, at the foothills of the Apennine mountains and within striking distance of the dramatic Hadley Rille. Apollo 15 was also the first of the so-called 'J'-type missions, in which the astronauts could motor around the surface in an electric powered Lunar Rover. (The unfortunate Apollo 14 astronauts were provided with a rather clumsy wheelbarrow affair to trundle their gear around in. This was not very dignified, but it was hailed incorrectly in the press as 'the first wheeled vehicle on the Moon'. The first wheels in fact belonged to the unmanned Luna 17 lander which reached the Moon in November 1970.)

A mild shock awaited scientists when the Apollo 16 mission returned. The site selection studies of Orbiter pictures had suggested that an area known as the Cayley Plains would prove to consist of volcanic rocks located within the lunar highlands. It turned out to be nothing of the sort: the rocks were not volcanic at all, but were much older highland material. The Taurus Littrow site was selected for Apollo 17 because it was thought to contain particularly young volcanic rocks, and again this proved to be not quite the case. In addition, both Apollos 16 and 17 were planned in the knowledge that the entire Apollo programme had been pruned; that Apollo 17 would be the last mission to the Moon for many years; conceivably the last *manned* mission for centuries. It was therefore particularly important that some series of surface experiments should be completed.

The surface experiments

The precious hours and minutes that the astronauts spent on the Moon were naturally meticulously planned. Never before in

Figure 3.13.   Site selection for the lunar landings aimed at obtaining the best value from each site. The Apollo 15 site (arrowed) *offered access to three major features: the surface of part of Mare Imbrium; a splendid sinuous rille (Hadley Rille); and the opportunity to sample material from the foothills of the Apennine mountains.* (Apollo Metric)

history can the movements of *any* human beings have been so carefully structured, or so rigorously monitored. In one sense, the astronauts were unimaginably remote from the Earth; so remote that the radio waves carrying their voices took over a second to leap across the space between Moon and Earth. In another sense, however, they were never alone for one second. At mission control in Houston a team of some of the best scientists and technicians in the world watched over them, monitoring their heartbeats even while they slept, and were always ready to advise them, to recalculate estimates, to suggest samples to collect, to modify schedules and to make sure that the astronauts did not stray too far either geographically or in time from the planning guidelines which ensured that they would be back inside the relative safety of their lunar module before their oxygen supply was exhausted.

The first task that faced the astronauts after they had emerged from their gawky, gold-foil wrapped lunar modules and practised a few bouncy steps on the lunar surface, was to deploy what was known as ALSEP, the Apollo Lunar Sciences Experiment Package. The package's contents varied slightly from mission to mission, but some nineteen major experiments were involved.

Seismic studies

Readers who have spent time in prison may have learned that, while it may be impossible to hear a conversation in the next cell if the walls are thick, banging on the wall, or, better still, on the central-heating pipes, provides an excellent means of transmitting information. The fact that rocks transmit acoustic shocks extremely

Figure 3.14. The Passive Seismic Experiment installed at the Apollo 11 landing site. This instrument provided the first direct data ever on the nature of the Moon's interior. (Apollo Hasselblad)

well underlies the whole science of seismology: the science of 'mapping with echoes'. The 'clang' produced by a nuclear explosion or earthquake taking place in, say, Kamchatka, travels in the form of shock waves right through the Earth and can be detected in any part of the Earth where the correct instruments are located. Studies of the ways in which these shock waves or seismic signals are bent and bounced around within the Earth have provided almost all the data that we have on the Earth's internal structure; the main evidence that the Earth has a liquid core comes from seismic studies. Seismic studies of the Moon's interior therefore formed one of the most important parts of the Apollo scientific programme.

Seismic stations were set up at five of the landing sites, and these were designed to go on working for long periods. The Apollo 11 station failed after a month, but the others continued working for years, providing an almost embarrassing amount of data. About 10,000 'moonquakes' had been detected before analysis of data coming in from the Moon was suspended. Most of these were very feeble, ranking at 1 or less on the Richter scale. (Earthquakes which cause damage or loss of life usually rank between 6 and 8.) About 1,800 meteorite impacts were also detected. (These, fortunately, have *never* been detected by seismic instruments on Earth. Arriving meteorites simply burn up in the atmosphere, for which we should be duly grateful.) The seismic signals produced by the impacts are quite different from those of the moonquakes.

The lunar seismic signals are also surprisingly different from terrestrial ones. Perhaps the most obvious difference is that they simply last longer. When the Saturn IV B rocket stages of the Apollo vehicles were deliberately crashed on to the Moon, their impacts set up seismic reverberations which lasted for many hours; on Earth, these would have lasted barely a few minutes. Moonquakes and meteorite impacts produced weaker signals, but these still lasted between thirty minutes and a couple of hours. It is thought that the explanation for this long period of reverberation lies in a heterogeneous layer a few hundred metres thick which covers the entire outer surface of the Moon. This layer has the property of scattering shock waves very efficiently *without* soaking up or 'attenuating' their energy. Effectively, the shock waves bounce around inside this layer until they gradually leak away down into lower levels of the Moon.

Many factors probably contributed to the properties of this layer, but the most important are undoubtedly the churning up caused by meteorite impacts, the total lack of water and strong bonds between individual rock particles. (The surfaces in contact

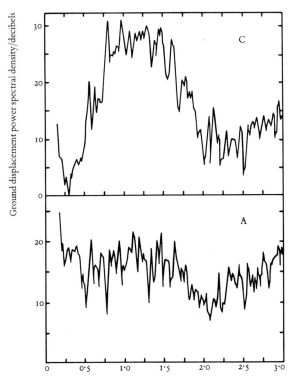

Figure 3.15. These graphs illustrate clearly the differences between seismic signals from meteorite impacts and genuine moonquakes.
The impacts (C) have a much more peaked curve than moonquakes (A).

are almost completely outgassed because of the extremely low atmospheric pressure, equivalent to a high vacuum on Earth, so that powerful bonds between individual grains are established.)

When the seismic network had been operating for over two years, an extraordinary fact was noticed: the apparent random 'scribbling' of seismometer traces for different moonquakes weeks apart matched each other precisely. Forty-one of these matching signals were discovered among the 600 largest signals. Clearly, the signals must have been produced by repetitions of the same kind of phenomenon. But what? And where?

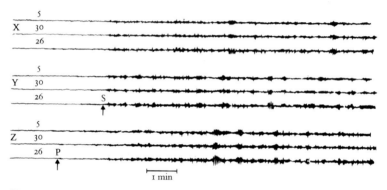

Figure 3.16.   The scribblings from Apollo seismometers illustrate the extraordinary 'matching' of moonquake seismic signals. Each trace relates to a different event; event 5 took place on 6 January 1970, event 30 on 23 May, and event 26 on 26 April.

It is quite easy to locate the source of an earthquake or moonquake. All one has to do is to time the arrival of each shock wave at each seismometer, and then, knowing the velocity of the shock waves, the difference in arrival times can be expressed in terms of difference in distance from the signal source. Given an adequate network of stations, the source can then be pinned down quite precisely. By carrying out this kind of calculation, it was found that the source regions of most moonquakes are all extremely deep, between 800 and 1,200 kilometres below surface, a remark-

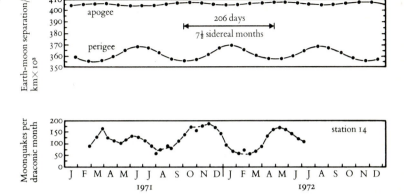

Figure 3.17. There is a well-defined correlation between variations in the shape of the Moon's orbit (upper diagram) *and the frequency of moonquakes* (lower diagram). *Moonquakes are most common when the Moon's orbit is most elliptical.*

able discovery in light of the fact that the vast majority of earthquakes take place at depths shallower than 500 kilometres.

This discovery showed that moonquakes are not caused by the same kind of shallow-level fault movements that cause earthquakes. When the *frequencies* of moonquakes were examined, another remarkable fact came to light. A fairly obvious periodicity can be seen in Figure 3.17; there is a peak in moonquake frequency every seven and a half months or so. This regularity suggested that moonquakes are a *tidal* phenomenon.

The shape of the Moon's orbit around the Earth alters slightly but regularly with the passage of time, which means that the minimum distance between Earth and Moon (at perigee) and the maximum distance (at apogee) also fluctuate. Figure 3.17 shows the monthly variations in the difference between perigee and apogee. The greatest frequency of moonquakes occurs at the times when the difference between perigee and apogee is greatest – in other words, during the period when the lunar orbit is most elliptical. (If there were no difference between perigee and apogee, the orbit would be circular.) The periods when the Moon's orbit is most elliptical are also the periods when the greatest tidal strains are set up within the Moon, because the variations in the gravitational force caused by the Earth on the Moon are then also at their greatest. The 'tides', of course, do not involve masses of water sloshing around on the surface. They affect the whole body of the Moon, and moonquakes are merely the expression of the very slight shifting around that takes place.

The overall structure of the Moon was gleaned from data from moonquakes, meteorite impacts and the impacts of nine discarded Saturn IV B rockets and lunar modules. It was found that the Moon has an outer crust some 60 kilometres thick, with masses of lavas filling the mare basins. Below the crust, a sharp transition takes place, comparable with the crust/mantle boundary on Earth, though this is usually much shallower; as little as 10 kilometres below the oceans. From geophysical data, the lunar mantle appears to be broadly similar to the Earth's. It extends down to a depth of about 1,000 kilometres, where another transition takes place. This transition is not fully understood yet, but it appears to mark the boundary of the Moon's core, which has a radius of 700 kilometres. Although the data are not nearly so clear as those for the Earth, there is some evidence that the Moon's core may also be liquid, or, at least, not solid. It is probably also very hot – a temperature of 1,500 °C has been suggested.

Whereas the study of moonquakes could tell seismologists a good deal about the deep interior of the Moon, the investigations of the structure of the outermost few kilometres required *active* rather than passive experiments to be carried out. In other words, the geophysicists had to devise some means of setting off the seismic shock waves. Readers will have seen this often enough on TV – most films of oil exploration crews feature the explosions on land or at sea that are used to produce shock waves.

On three of the Apollo missions, similar techniques were used. On Apollo 17 a series of eight explosive charges were laid out, and then fired on a command from Earth. In addition, the discarded lunar module was crashed only 8 kilometres from the landing site, producing a fresh supply of signals. These experiments

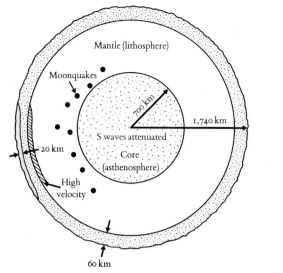

Figure 3.18. *A cartoon illustrating the overall structure of the Moon, with core, mantle and crust. The distribution of mare basins and moonquake centres is shown schematically.*

showed that, in the Taurus–Littrow valley, there is a layer of bedded lavas and broken rocks some 270 metres thick, resting on top of a layer of solid lavas over 1 kilometre thick. These data helped to refute some lingering doubts that still existed in the minds of those who had argued that the mare basins were full of dust, increasingly compacted with depth!

Thermal studies

Making a note of the temperature on the surface of the Moon might seem an obvious measurement for the Apollo astronauts to make. But it was not as simple as that. When one talks about the 'temperature' on Earth on a hot sunny day, what one actually means is the *air* temperature. But there is *no* air on the Moon, and in some senses therefore the concept of temperature does not exist there. It is, however, possible to measure the temperature of the surface materials. This was found to vary from a maximum of about 111 °C to a minimum of −171 °C, figures closely corresponding to those that had been derived by remote sensing studies.

A much more important measurement for geological purposes was that of the *heat flow*, which is basically a measure of the Moon's own *internal* heat and gives an indication of the abundance of radioactive elements (which produce the heat) and of the temperature in the interior. Two misfortunes hit the planned series of experiments. The first instrument was carried on the ill-fated Apollo 13 mission, which failed to land. No instrument was carried on Apollo 14, but there was one on Apollo 15, and a successful measurement was made in the Hadley Rille area. Apollo 16 also carried an instrument, and its results were eagerly awaited, since it would yield data from a contrasting highland region. Unfortunately, in laying out the ALSEP package on the Moon, one of the Apollo 16 astronauts tripped over a critical cable and the experiment was ruined. Accidents can happen anywhere! The only other result obtained was from the Apollo 17 Taurus-Littrow suite.

Two determinations are not many for a body the size of the Moon, but both were closely similar at about 0·7 heat flow units, so this figure is probably representative of the Moon. The figure is significant because it is about half the average heat flow of the Earth, and because it demonstrates unequivocally that the abundances of radioactive heat-producing elements, such as uranium, potassium and thorium, are far too high for the Moon as a whole to have the composition of a *chondrite*, a class of meteorite which some scientists had considered represents the primitive material from which the Moon might have been formed.

Magnetometry

The Moon's magnetic field is exceedingly complicated, but it has some important implications. Loosely speaking, three types of field are present: (a) the external field (which was measured by the orbital experiments), (b) permanent fields contained in the rocks at the landing sites and (c) fields produced by interactions with the solar wind.

The solar wind is an important factor in many lunar studies, not only those connected with magnetism. It is, of course, nothing like a terrestrial breeze. It is difficult to define precisely what it is, but one approximation might be: a continuous flow of ionized gas (or plasma) streaming out spirally from the Sun's atmosphere, mainly in the form of protons, electrons and complete hydrogen atoms, all moving at velocities of hundreds of kilometres per second. It is, in effect, gas escaping from solar gravity and expanding into the vacuum of space.

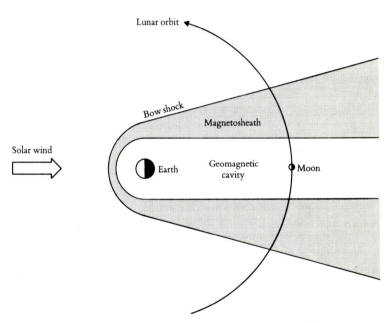

Figure 3.19. The interaction between the Earth's magnetic field and the solar wind. These have to be understood and taken into account before the Moon's own magnetic field can be interpreted. The lunar magnetic field is quite different where it is screened by the Earth from where it receives the full blast of the solar wind.

As the solar wind flows through the solar system, it interacts with comets, the Moon, planets, interplanetary dust and cosmic rays. If a planet possesses a magnetic field, the solar wind is arrested on the sunward side of the planet at a distance such that the dynamic pressure of the wind is equal to the opposing 'pressure' of the compressed magnetic field of the planet. A 'bow shock' and a cavity are produced somewhat analogous to the shock waves developed by a supersonic aircraft.

The interaction between the solar wind and the Moon is quite different since, as mentioned earlier, the Moon has such a weak magnetic field. The solar wind particles reach the surface, and are absorbed there. This absorption creates a 'hole' (called the *diamagnetic lunar cavity*) in the solar wind, and there is no lunar 'bow shock'.

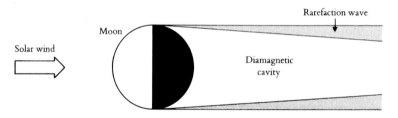

Figure 3.20.   Because of its weak magnetic field, the Moon interacts quite differently with the solar wind, compared with the Earth. A 'hole' or shadow, called the diamagnetic lunar cavity, *is created.*

The complications caused by the solar wind in interpreting the Moon's magnetic field are least when the Moon is passing through the Earth's geomagnetic tail, when the Moon is screened by the Earth. It is during these periods that measurements can best be made. Perhaps the most important discovery was that the ancient magnetic fields contained in the rocks of the Moon – known as the natural remnant magnetism – are much stronger than the present external field, by a factor of 60 in some cases. Studies on some of these rocks suggest that, when they acquired their magnetism, the Moon's external field must have been much stronger, perhaps as much as one twentieth of the Earth's field.

All this is a long way from everyday experience. Why on Earth, one might ask, is the Moon's magnetic field of the remotest interest to anyone? The reason is that the existence of a magnetic field is closely connected with the internal structure of a planet, and provides one of the few ways of investigating that structure. In the case of the Earth, it is thought that its powerful magnetic field can only be generated by an immense natural dynamo operating in the core, where swirling currents of molten metal interact with one another. The *absence* of a strong field in the Moon suggests that it does not have a molten metallic core – as is borne out by density studies – but it could possibly have had a liquid core of some kind when the natural remnant magnetism was acquired by the surface rocks. Another explanation is that a 'permanent' magnetic field was imprinted on the Moon by some external agency. If, as seems certain, the Moon was at one time much nearer the Earth, the Earth's magnetic field could have been responsible. Alternatively, it is possible that, early in the history of the solar system, the Sun's magnetic field may have been up to a hundred times stronger than it is now.

Long-range surveying:
the Laser Ranging Retroreflectors

These instruments – called LR^{3}s for short – were located at three Apollo sites. Basically, they are nothing more than sophisticated reflecting prisms, capable of reflecting back pulses of laser light sent from Earth. They are wholly passive, so in theory they should be usable indefinitely. The existence of these reflectors on the Moon enables the Earth–Moon distance to be measured extremely precisely, simply by measuring the interval between the start of a laser pulse and the arrival of the reflected pulse. This makes it possible to improve the accuracy of our knowledge of the Moon's motion and libration, and also of the Earth's own rotation.

If observations are continued over several years, these reflectors may also enable us to keep tabs on continental drift. Although it is possible to demonstrate on Earth that one continent is moving *relative* to another – America is moving away from Europe, for example – it is only by relating this apparent motion to a third, fixed point that the true motions of the two continents can be established.

Sun-tanned rocks, the solar wind and cosmic rays

The solar wind, it was explained earlier, consists of atomic particles blasted out of the Sun at enormously high velocities. Most of the particles consist of hydrogen and helium, with fairly low energies: the solar wind is usually a steady, gentle stream. Sometimes, however, great flares erupt on the surface of the Sun, and much higher-energy particles are sprayed out into space. Even these, however, are nowhere near as energetic as cosmic rays, which travel at velocities near the speed of light. These rays consist of atomic particles of much heavier elements than those in the solar wind, but exactly how and where these powerful rays are produced is not fully understood. It is certain that they do not come from the Sun, and must come from much more distant sources in deep space, perhaps from regions in the Milky Way galaxy, or maybe from even further out. An individual particle in a cosmic ray is almost unimaginably small, yet it may be travelling so fast that it has as much energy as a tennis ball smashed in a service ace by a Wimbledon champion.

The Apollo studies were concerned with investigating the nature and origin of the atomic particles making up the solar wind and cosmic rays, and with examining the results of the prolonged exposure of moon rocks to their searing effects. For both these studies, the Moon provided a convenient orbiting space station; an excellent platform on which to study processes taking place throughout the solar system.

The astronauts themselves proved – unwittingly at first – to be quite sensitive cosmic ray detectors. Rays passing through their heads and eyes caused them to see occasional bright flashes, even with their eyes shut. Although it may not seem too pleasant to have atomic particles streaking through one's body at speeds near the velocity of light, the astronauts suffered no ill effects from their experiences. On longer space flights, however, the accumulated damage to body tissue caused by the particles could present a more serious problem.

While on the Moon, the astronauts were bathed in the full glare of radiation from solar and galactic sources. Quantitative studies of the atomic particles hitting them were made by examining their helmets. Each high-speed particle striking a helmet bored deeply into it, producing a microscopic track which could be revealed by acid-etching techniques on Earth. Up to three tracks per square centimetre were found, indicative of the number of particles that had passed through the astronauts' heads! A slightly more sophisticated version of this was designated the Solar Wind Composition Experiment (SWCE), which consisted simply of hanging a bit of aluminium foil from a pole stuck into the lunar surface immediately after landing, and collecting it again before leaving. The cosmic ray experiments were similar, and involved leaving chunks of materials such as aluminium, mica, platinum and glass exposed on the surface of the Moon during the missions. In some cases, the specimens were placed facing the Sun, so that comparisons could be made with identical specimens facing *away* from the Sun, so that the numbers of particles arriving from each direction could be measured separately.

On the Apollo 12 mission, some pieces of the Surveyor 3 craft which had been exposed for over two years were also collected. No evidence of any impacts by micrometeorites – ordinary, solid particles – was found, but acid etching provided many data on the speeds, directions and energies of the cosmic ray particles which had entered the spacecraft in the two years it had stood on the Moon.

The extraordinarily high energies of solar wind and cosmic ray particles have some useful side-effects. Solar wind particles travel so fast that they can penetrate the surface layers of rocks, but cosmic ray particles are so energetic that they can pass through a metre or more of solid rock. Both kinds of particle damage the crystal structures of the minerals they pass through,

leaving tracks that can be detected by acid etching and microscopic examination. The study of these tracks has provided a
unique way of determining the ages of the rocks on the surface
of the Moon: just as one can get an idea of how long a sunbather
has spent on a beach from how brown – or sun-burned – he is,
so one can get an idea of how long a rock has spent on the lunar
surface from how many cosmic ray tracks it has within it.

Making the assumption that the number of cosmic ray particles
arriving has been constant through time, all that is required is to

*Figure 3.21.   Suspended from the pole in front of the astronaut is a small
piece of aluminium foil: this represented the collecting device for the Apollo
11 Solar Wind Composition Experiment.* (Apollo Hasselblad)

count the number of tracks in a given area of a mineral crystal. Even better, since different kinds of particles penetrate to different depths below the surface, it is possible to determine how long the specimen has spent at different depths, and how long it has spent on the surface itself. Not surprisingly, the term which is used to describe the time a rock has spent on the surface is known as its 'sun tan' age.

These techniques required examination of returned samples, so we are to some extent anticipating the next section. The results, however, provided such a stimulating insight into lunar surface processes that it would be tedious to delay it. Track studies showed that most specimens had 'sun tan' ages of less than three million years, indicating that they had actually been exposed on the surface for this time. Most specimens, however, seem to have spent no less than *100 million* years within 10 centimetres of the surface! Such a thing, of course, could never happen on the Earth. A hundred million years ago, dinosaurs still roamed around and mammals had scarcely evolved. Since that time, the Earth's surface has changed out of all recognition. Great mountain ranges have reared up, and others have been worn away. Some oceans have opened, and others have closed. An Ice Age has come and gone.

On the Moon, nothing whatever happened during all those millions of years, save for gentle 'gardening' of the surface by micrometeorite impacts, which occasionally flipped rocks over or brought them to the surface, or covered them with a thin layer of ejected material. This stirring up of the lunar surface appears to have affected only the topmost few centimetres; deeper stirring is much rarer.

A further factor which these studies revealed is that the surfaces of rocks exposed on the Moon are physically eroded away at a rate of less than 1 millimetre per million years by micrometeorite bombardment and by *sputtering*, the interaction of solar wind and cosmic ray particles with the atoms making up the rock. This is perhaps the single most important process actively taking place on the Moon. The discovery is significant, because other bodies in the solar system are exposed to exactly the same kind of process. They range in size from the planet Mercury to the tiniest micrometeorites. Clearly, small particles which are close enough to the Sun to have sputtering rates close to 1 millimetre per million years will have strictly finite lives.

The returned samples

A total of 381·69 kilograms of rock samples was collected by the six Apollo missions. Since the Apollo project as a whole cost some $40 billion, one is tempted to say that these rock samples, the only *tangible* results of the lunar landings, are worth $100 million per kilogram. At that price, they are worth many times their own weight in gold – or even diamonds. Such financial discussions are not very useful, though. The simple fact is that the specimens are effectively irreplaceable, and therefore NASA guards them jealously. A special building, the Lunar Receiving Laboratory, was built to house them at Houston, and there most of them remain. Some of them are still maintained in the same conditions in which they were collected. Although a lot of money was poured into the Apollo project in the early days, this did not remain the case, and at one stage disgruntled scientists could be heard speculating bitterly on the point of spending enormous sums on collecting lunar samples, and then not being able to find enough money to keep the roof of the Lunar Receiving Laboratory from leaking in wet weather!

Whatever NASA's financial problems, there can be no doubt of the enormous value of the scientific information obtained from the lunar specimens. Samples of the material were sent to distinguished scientists all over the world for detailed investigations. The volume of data generated from this work is astronomical. More is being published each year. There is even a special journal dealing exclusively with the Moon. The situation has now arisen where, in some areas of study, more data are available for lunar material than for the terrestrial equivalent. A bewildering range of investigations were carried out, far more extensive than can be summarized here.

The moon rocks

When the Apollo missions were being planned, there was a good deal of anxious debate about the horrendous consequences for the Earth if the astronauts were to bring back with them exotic moon bugs or bacteria that might escape and multiply on Earth. Many ghastly predictions were made of people toppling like ninepins as they succumbed to epidemics of strange new diseases sweeping round the world. To prevent this particular version of biological doomsday from coming about, elaborate steps were taken to quarantine the Apollo crews and their samples on return to Earth. This was to a large extent a public-relations exercise, because the first thing that the Apollo 11 crew had to do when they got back, of course, was to open the hatch of their space-craft. The spacecraft was heavily contaminated with lunar dust brought in from the Moon on their spacesuits, and any bugs living in the dust would immediately have been wafted through the hatch on balmy Pacific breezes and have started multiplying. The Pacific Ocean might even have provided them with an idyllic bath of warm, nutrient fluid.

We all know now, of course, that the Earth escaped a cata-strophic plague. When the lunar samples were examined, they were found to show no trace of biological activity whatever; they were about as inert and sterile as it is possible for any material to be. Far more realistic quarantine procedures, however, will be needed when samples are brought back from other parts of the solar system, including Mars. Conditions on the Moon are such that it was a near certainty that nothing could live there; the odds are much less heavily stacked against the existence of microbes on Mars.

When the first bags of samples brought back by Apollo 11 were opened, there was naturally a good deal of excitement among the waiting scientists. At first sight, however, there was not apparently much to get excited about. Most of the bags contained drab, dusty, grey-looking material, about as stimulating to look at as clinker from a boiler. Closer examination revealed that the specimens were a good deal more interesting than they at first seemed. Several different types of specimen were present: regolith material, glasses, breccias and solid rocks.

The *regolith* material can be thought of as the lunar 'soil'; it consists of fine-grained dusty particles, produced by the churning up of the outer surface of the Moon by meteorite impacts. Its existence came as no surprise – in fact, it was exactly what most scientists expected. But the extremely adhesive nature of the dust was not fully anticipated, and it caused a number of problems to the astronauts, clogging up their suits and jamming some pieces of equipment.

Glasses are an important part of the regolith. The glasses were formed when meteorite impacts took place, the energy of the impact both melting the rocks at the impact site and hurling droplets of the molten material huge distances over the Moon. The glass particles are tiny, abundant and easy to analyse; they thus provide splendid samples from areas of the Moon not visited by astronauts. Their compositions give a guide to the range of rock types on the Moon; and the abundance of particles of different compositions gives a guide to the relative abundances of the source rocks. It is not possible, though, to identify the source region from which any individual particle came.

Figure 3.22.   This Apollo 11 bootprint conveys clearly the friable but cohesive nature of the surface of the regolith. Fresh, powdery snow on Earth is superficially similar. (Apollo Hasselblad)

The glassy material occurs either as agglutinate, a glassy film cementing together other fragments of rock, minerals or older glass, or as smoothly rounded spheres and dumb-bells. The spheres are among the most frequently photographed lunar rock specimens, since they are beautifully spherical, and although they may only be a fraction of a millimetre in diameter, they often show perfect micrometeorite craters. These have been called rather suitably 'zap pits', and they demonstrate convincingly just how intense the micrometeorite bombardment of the surface is.

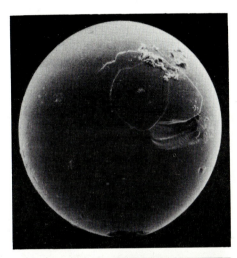

Figure 3.23.   A green glass sphere from the 'soil' at the Apollo 15 site. The sphere is about one tenth of a millimetre in diameter, and part of it has been chipped away by an impact. (National Space Science Data Center)

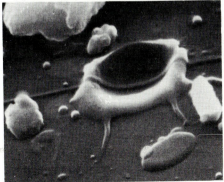

Figure 3.24.   A 'zap pit' about one tenth of a millimetre in diameter caused by a micrometeorite impact on an Apollo 15 rock specimen. (National Space Science Data Center)

Another product of meteorite impacts are the *breccias*. These
are rocks which themselves consist of angular fragments of other
rocks welded together in a glassy matrix. The fragments were
produced by the terrific shock waves generated by large crater-
forming impacts. So intense were these disturbances that most of
the breccia samples show evidence of *shock metamorphism*. The
impacts involved the release of such huge amounts of energy that
the pressures set up in the shock waves instantaneously reached
thousands of times the Earth's atmospheric pressure, and the
temperature thousands of degrees. Although transient, these
effects were sufficient to affect the rocks profoundly, so that

*Figure 3.25.   An enormous boulder from which geological samples were
collected during the Apollo 17 mission. The rock making up the boulder
was a breccia, a mixture of angular fragments welded together.* (Apollo
Hasselblad)

their original nature was barely recognizable. In some cases, the shock-metamorphosed fragments in a breccia sample turned out themselves to be brecciated, having already been torn apart and welded together again by an earlier impact event.

Much the easiest samples to deal with were the *solid rocks*, because they could readily be identified and interpreted by earth-bound geologists. Two main types were found, one of them familiar to every geologist, the other much less so. The familiar type was *basalt*, which on Earth makes up all of the ocean floors and is erupted by most active volcanoes. On the Moon, all the maria were found to be made of basalt lavas, as had confidently

Figure 3.26. A close-up of a lunar breccia. This one was collected on the Apollo 16 mission to the lunar highlands. The fragmental texture is very clear, as is the pale colour of the matrix. (National Space Science Data Center)

been predicted. The unexpected rock type was *anorthosite*, a pale-coloured rock which is not common on Earth and is found only in ancient parts of the continents. Almost all the light-coloured highland areas of the Moon appear to be made of anorthosite, or related rocks. This was something of a surprise, because, although geologists were aware that the highlands were certainly older and different from the maria, in pre-Apollo days anorthosite did not initially seem to be a likely candidate – in fact, it was never even mentioned. It was subsequently found, though, that anorthosite has played a major part in the evolution of the Moon.

Lunar minerals

Those science-fiction addicts who hoped to find on the Moon inexhaustible supplies of gold, diamonds and other desirable minerals must have been mildly disappointed by the Apollo samples. The range of minerals found on the Moon is rather limited, with far fewer individual mineral types being found there than are known on Earth. There are sound geochemical reasons why no geologist should seriously have expected to find abundant gold and diamonds (although the latter were seriously predicted by at least one scientist), but the small number of different minerals can be more simply explained.

The commonest minerals on the Moon are rather ordinary ones which are also common on Earth: feldspars, pyroxenes, olivines and various oxides. These are the minerals which make up common igneous rocks like basalt. What is missing on the Moon is any kind of *hydrous* mineral, one that contains water in its atomic structure. Such minerals are extremely abundant on Earth. Their absence on the Moon indicates that the Moon is, and has always been, an exceptionally dry place. The presence of native iron in many lunar specimens also demonstrates another important point: the Moon is, and must always have been, an oxygen-poor environment. (Iron in terrestrial rocks would rapidly be oxidized to iron oxides, just as iron cars exposed to the atmosphere are rapidly oxidized to expensive rust-coloured scrap.)

Only a few completely new minerals were found on the Moon, fascinating to mineralogists, but otherwise rather obscure. One of them, an oxide of iron, magnesium and titanium, was named *Armalcolite*, after the Apollo 11 astronauts, *Arm*strong, *Al*drin and *Col*lins.

The chemistry of the lunar rocks

An amazing range of chemical analyses were made of the returned samples. Since a total of only 400 kilograms was available, it was essential for the analysts to use as little as possible of the material, and consequently some highly sophisticated analytical techniques were developed. In one case, where lead isotopes were being studied, the amounts of material available were so small, and the detection limits so low, that sources of contamination like ordinary electric light bulbs had to be eliminated. Believe it or not, the lead vaporized from the silvery-grey contacts at the base of the bulb represented a possible source of contamination! The laboratory in which these studies were made was a miracle of scrupulous cleanliness – it was far cleaner than the average hospital operating theatre.

Many important discoveries concerning the origin and evolution of the Moon and the solar system were made as a result of these analyses. They provide the key to many solar system studies, so they are described fairly fully here.

The mare basalts

All the earlier work had suggested that the huge maria basins were filled with many thin basaltic lava flows, and the returned samples showed that these basalts did indeed look strikingly similar to ordinary terrestrial basalts. Chemically, however, there are some important differences. The lunar basalts are much richer in iron and titanium than terrestrial ones, but poorer in alkali metals, such as potassium and sodium. Some even more important differences are revealed by trace elements (elements which occur

Figure 3.27. An Apollo orbital photograph of part of the Mare Imbrium, showing the crater Euler and an isolated mountain called La Hire. To the left of La Hire, tongues of basalt lava can be seen snaking over the surface of the mare. The flows were probably erupted 3·2 billion years ago. (Apollo Metric)

in only minute amounts, of the order of parts per million). These differences are summarized graphically in Figure 3.28. It is clear that the lunar basalts are enriched in chromium, but poorer in potassium, sodium, bismuth, gold and iridium.

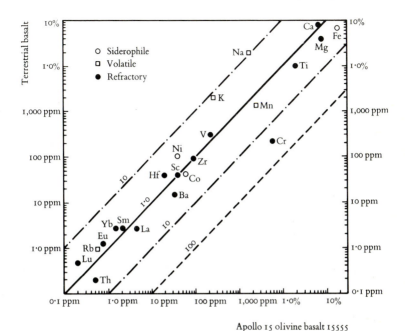

Apollo 15 olivine basalt 15555

Figure 3.28. A graphical comparison of the chemical compositions of a lunar and a terrestrial basalt. Elements which are more abundant in terrestrial basalts fall to the left of the bold line; elements which are more abundant in lunar basalts plot to the right. Elements plotting along the line are equally abundant; the further an element plots away from the line, the greater the contrast between terrestrial and lunar basalts. (From S. R. Taylor, Lunar Science: A Post-Apollo View, Pergamon Press, 1975)

These data show that lunar basalts are quite dissimilar from terrestrial ones and, by implication, could not have been formed from the same source. This is important, because one long-standing argument had been that the Moon might have formed initially by splitting off from the Earth, and that the Pacific Ocean represents the enormous scar left by the fission process. If that had been the case, one would have expected the interior of the Moon to be of the same composition as the interior of the Earth, and therefore that basalt lavas on both Moon and Earth should show strong affinities. They do not.

A second and much more fundamental point concerns the links between the composition of the Moon and the origin of the solar system. It has long been thought that a certain type of meteorite, known as a CI carbonaceous chondrite, has a composition closely similar to that of the primitive solar nebula, the cloud of gas and dust from which the planets formed. These meteorites may, in fact, be nothing more than chunks of scrap material left behind in space after the planets formed.

Now, when the trace element compositions of lunar basalts and CI chondrite meteorites are compared, some important differences emerge. The lunar basalts are rich in *refractory* elements, such as calcium, aluminium, titanium and zirconium, but poor in *volatile* elements, such as sodium and potassium. (A refractory element is one that melts and vaporizes at extremely high temperatures. Titanium requires temperatures over 3,000 °C to vaporize it, sodium only 883 °C.)

An important conclusion can be drawn from these differences. The lunar basalts could *not* have been derived from a source which was of straightforward CI chondrite composition. In other words, the lunar interior could not be made of primitive solar nebula material, as some scientists had speculated. Furthermore, the Moon could not be a source of CI chondrites, as some scientists had also hoped. Thus extensive fractionation or splitting up of the elements was involved in the formation of the Moon from the solar nebula. This is a vital point, which will be important in considering the origin of the solar system as a whole.

The highland rocks

The highland rocks are much older and more difficult to interpret than the mare basalts. To quote a distinguished author on lunar work, S. R. Taylor:

The meteoritic bombardment of this [highland] surface has drawn a curtain across the landscape through which we peer dimly to discern an earlier history. The oldest landscape is saturated with large craters and probably with giant ringed basins. Thus the earlier structure of the lunar crust is obliterated and there is 'no vestige of a beginning'.*

Most of the difficulties in interpreting the highland rocks stem from the extensive shock metamorphism and brecciation of the rocks. Chemical work showed that a number of different rock types are present, and a large number of acronyms was spawned to describe them: ANT, ANTIC, KREEP, META-ANT and so on. For our purposes, however, we can think of the most important highland rock type as anorthosite, which consists almost entirely of pale-coloured plagioclase feldspars.

Anorthosites, although not widely known on Earth, are among the most infamous lunar rocks. This notoriety stems from the discovery during the Apollo 15 mission of a white lump of anorthosite perched upon a lump of regolith dust. It was hailed immediately in the press as the 'Genesis Rock', because the Apollo scientists were unwise enough to suggest that the first sizeable lump of highland rock that had been obtained might hold the key to the *petro*-genesis of all lunar rocks, meaning the origin of the rock groups. Their modest hopes were, of course, blown up out of all proportion by the press, and nearly every newspaper in the world carried a photo of the 'Genesis Rock'.

The trace element composition of the anorthosite shows roughly the same variations relative to C1 chondrites as the basalts. There is one interesting difference between the anorthosites and basalts, however, concerned with the single element europium. Europium is a fascinating element, which has the useful property of slotting neatly into the atomic structure of feldspar minerals.

Imagine a molten magma from which feldspar is crystallizing. If europium preferentially concentrates in the feldspar, then, when a large amount of feldspar has crystallized and sunk to the bottom of the hypothetical melt, the feldspars will have more than their fair share of europium, and the residual liquid less than its fair share. The analytical data showed that the mare basalts are

* S. R. Taylor, *Lunar Science: A Post-Apollo View*, Pergamon Press Inc., New York, 1975.

Figure 3.29. The 'Genesis Rock' as it was found by Apollo 15 astronauts, perched up on a small pinnacle of regolith material. The tripod-like object is a gnomon, used to provide a scale and a colour-test chart in lunar photos. (Apollo Hasselblad)

characteristically poor in europium, but the anorthosites, which consist virtually of solid feldspar, are *rich* in europium, so they have more than their fair share. This suggests that the europium-poor basalts were derived from materials *from which the feldspars making up the anorthosites had already been subtracted*. This is another factor to bear in mind when considering the evolution of the Moon.

The age of the Moon

The age of the Moon is a subject that we have skated round so far, although the 'sun tan' ages of the surface samples were a clear pointer to the fact that the Moon has had an extremely long history. Scientists struggled for many years to work out the age of the Earth. It was clear that it had *not* been created at 9 a.m. on 26 October, in the year 4004 B.C., as had been suggested by Archbishop Usher from exhaustive biblical studies, but it was extremely difficult to come up with alternative *quantitative* estimates. One of the best was made by the distinguished physicist Lord Kelvin, who tried to calculate an age from the present temperature and heat flow, assuming that the Earth had formed at high temperatures. His estimate of about 30 million years was way off the mark, and was founded on an incorrect premise, but it was accepted for many years.

The breakthrough came with the discovery of radioactivity, and the realization that radioactive isotopes decay at an absolutely constant rate to produce other easily detected radiogenic isotopes. Thus, if the amount of a naturally occurring radiogenic isotope present in a rock *now* can be measured, the *age* of the rock can be found, using the rate of decay like the flow of sand in an egg-timer. These radiometric techniques have shown that the Earth, and all the meteorites that have been examined, were formed about 4·6 billion years ago. This date is taken to be the age of formation of the solar system as a whole.

The radiometric ages obtained for lunar rocks are much the most interesting and unexpected discoveries made throughout the Apollo missions. Prior to Apollo, scientists had only hazy ideas about the age of the rocks on the Moon. There was a general feeling that the surface was rather old, but that was as far as it went. Crater-density studies could be used to work out *relative* ages of lunar formations, but there was no way of calibrating these observations to get *absolute* ages – ages expressed in years.

The first samples of Apollo 11 mare basalts to be dated startled the scientific community – they were *3·8 billion years old*. This was quite unexpected. Terrestrial geologists are used to talking in terms of millions of years, and to dealing with ancient rocks, but rocks older than three *billion* years are exceedingly scarce on Earth. And, of course, it was quite obvious from crater density studies that the lavas of the Sea of Tranquillity *are among the youngest rocks on the Moon!*

This unexpected result was soon confirmed by dates on samples returned from the other mare sites. In fact, it soon became clear that almost *all* the lunar basalts were erupted between 3·8 and 3·2 billion years ago. This discovery has enormous implications. It means that the Moon has been effectively 'dead' since 3·2 billion years ago, save for the formation of a few impact craters. It means that the Moon has been a sterile, unchanging place throughout almost all the eventful geological history of the Earth. It also means that the chances of there being any present-day volcanic activity – as some moonwatchers had hoped – are exceedingly slim. (To keep things in perspective, remember that the first 'ordinary' fossils did not evolve on Earth until 600 million years ago, by which time the Moon had already been 'dead' for 2·6 *billion* years, and that continental drift did not operate to split off America from Europe until a mere 170 million years ago.)

Dated samples from the lunar highlands gave even older dates, most of them over 4 billion years. One sample even yielded a date of 4·6 billion years. This rock was clearly as old as anything in the solar system, and must have formed part of the very first crust of the Moon.

Certain other even more sophisticated studies were able to show that some enormous cataclysmic event – or series of events – had affected the Moon prior to 4 billion years ago. The effects of these events were so far reaching that they 'reset' the radioactive clocks in all the lunar rocks. One of the teams working on this project described themselves as the 'Lunatic Asylum of the Charles Arms Laboratory'. Their lighthearted name for themselves belied the seriousness of the work they did. By using some exceptionally brilliant analytical techniques, they were able to show that the Moon had experienced what they called a 'terminal lunar cataclysm' about 4 billion years ago.

What could this cataclysm have been? The answer was not difficult to find: it must have been a series of enormous asteroid impacts on the Moon, which sculpted the large craters and ring basins. This view is now widely accepted by most lunar scientists, except that it is generally thought that, instead of there being a single 'cataclysm', there may have been several, or even a 'continuous cataclysm'. Regardless of the number of cataclysms, it is clear that the intensity of the asteroid bombardment dropped off abruptly after 4 billion years ago (Figure 3.30).

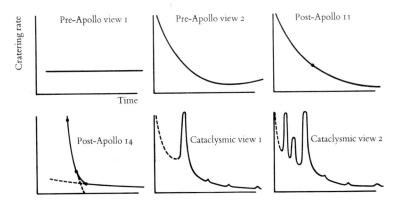

Figure 3.30.   The evolution in ideas about the cratering history of the Moon. The earliest workers thought that the Moon had been exposed to a constant rate of bombardment throughout its history. As the Apollo missions proceeded, it became clear that most of the cratering had taken place very early on, and that there may have been one or more 'cataclysmic' episodes.

The origin and evolution of the Moon

The origin of the Moon is a subject that is at one and the same time eternally fascinating and eternally frustrating. Fascinating, because it is always stimulating and mind-stretching to think about the origin of any part of the solar system; frustrating, because it is never possible to be sure of anything. Three main ideas about the origin of the Moon were in circulation prior to Apollo, and the same three appear to have escaped almost unscathed from the terrific blast of scientific work that has been aimed at them. It is fair to say that the origin of the Moon is nearly as much in doubt now as it was prior to Apollo, but we have learned more about what could *not* have happened.

One of the earliest theories was the *fission* theory, usually attributed to George Darwin, son of the great biologist. This argues that the Moon split off from the Earth very early on in history, a later refinement suggesting that the Pacific Ocean was the scar left by the fission. Then the *double planet* theory argued that the Moon is not really a moon at all – meaning that it is not a true satellite of the Earth – but rather a separate small planet which formed from the primitive solar nebula at the same time and at roughly the same point in space as the Earth. Both the 'fission' and the 'double planet' theories, however, seem improbable in the light of the chemical data, which show significant differences in composition of the Earth and Moon. If either theory had been correct, much greater similarities in composition would be expected. Both, however, still have advocates in favour of them.

The third theory is the *capture* theory. It argues that the Moon is a small, separate planet formed from a different part of the primitive solar nebula from the Earth, and later gravitationally 'captured' by the Earth. This overcomes some of the problems posed by the chemical differences, but the mechanics of the actual 'capture' process have yet to be plausibly explained. This theory is perhaps the most popular at present.

Our knowledge of the evolution of the Moon is more water-tight than that of its origin, because there are so many more constraints. Any discussion of the evolution of the Moon must be able to explain the internal structure, the absence of a magnetic field today but the existence of one earlier on, the chemical compositions of lunar basalts and highland rocks, the ages of the rocks, and so on.

The story starts way back in the mists of time – or, more literally – the dusts of time, 4·6 billion years ago when the solar system was born from a vast cloud of dust and gas. The Moon itself started off as a myriad of dust particles which came together or *accreted* into a spherical lump. Already an important step had been taken: the ball of homogeneous dust that was the proto-

Moon was different in composition from the bulk composition of the solar nebula. It was enriched in the refractory elements, but depleted in volatile elements, presumably because the high temperature did not allow for the condensation of the more volatile elements.

The process of accretion is supposed to have taken place extremely rapidly, at least on the time scale of the solar system. It may have taken place in a period of less than a million years, perhaps of only 100,000 years. Enormous amounts of energy were liberated by the myriads of collisions involved in the accretion process. The cumulative effect of the release of this energy in such a short space of time was that a large part of the primitive Moon melted completely. The outermost 1,000 kilometres may have been molten for a short while. A chilled skin probably formed rapidly on the extreme surface, but this was repeatedly disrupted by continuing major impacts, which represented the tailing-off of the accretion process, and fragments of the skin may be present in the lunar highlands, accounting for the exceedingly ancient rock specimens found there.

Within the liquid part of the Moon, dense crystals began to settle towards the centre, piling up deep within the lunar mantle, while lighter ones floated upwards to concentrate beneath the chilled surface. These crystals were mostly feldspars; ultimately, they accumulated to form the first crust of the Moon, some 60 kilometres thick. Below this crust denser crystals accumulated, and a general sorting out of chemical elements took place. Thus, as a direct result of the accretional melting, the Moon acquired a well-defined layering. This may have been established as early as 4·5 billion years ago, but the outer crust continued to be disrupted by cataclysmic impacts until about 3·9 billion years ago.

These cataclysms were responsible for the battered, scarred highland surface of the Moon that we see today, and, in particular, for the formation of the huge ring-basins. Initially, these basins were *not* filled with the smooth dark lavas that now cover them. The first basalts were not erupted until well *after* the main impacts had taken place. It is thought that the earliest basalts were produced by melting of lunar mantle materials at relatively shallow

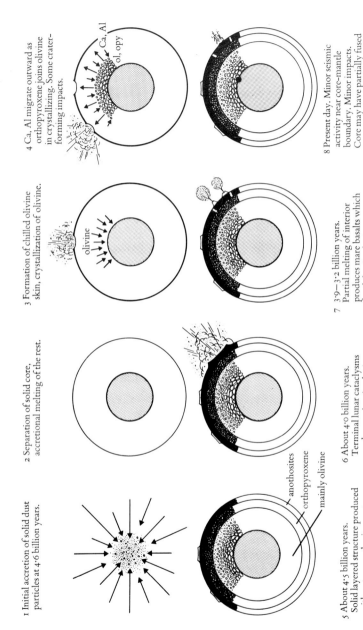

1 Initial accretion of solid dust particles at 4·6 billion years.

2 Separation of solid core, accretional melting of the rest.

3 Formation of chilled olivine skin, crystallization of olivine.

olivine

4 Ca, Al migrate outward as orthopyroxene joins olivine in crystallizing. Some crater-forming impacts.

Ca, Al
ol, opy

5 About 4·5 billion years. Solid layered structure produced with outer anorthosite crust.

anothosites
orthopyroxene
mainly olivine

6 About 4·0 billion years. Terminal lunar cataclysms produce major ring basins.

7 3·9–3·2 billion years. Partial melting of interior produces mare basalts which flood the ring basins.

8 Present day. Minor seismic activity near core-mantle boundary. Minor impacts. Core may have partially fused owing to radiogenic heating.

Figure 3.31. A cartoon summarizing the evolution of the Moon.

levels; later, the melting zone got deeper and deeper. This melting affected only small volumes of material, and, unlike the first large-scale (accretional) melting, it probably occurred in response to slow accumulation of heat from the decay of radioactive elements.

About 3·2 billion years ago, the last lavas were erupted. These seeped to the surface from great depths within the Moon; possibly over 150 kilometres. Subsequently, no melts were able to make the long journey to the surface, and lunar volcanism fizzled out.

For hundreds of millions of years, the Moon has remained substantially unchanged. Had anyone lived on Earth 2 billion years ago, the Moon he saw then would look almost exactly the same as the one we see today. The surface has been moulded over the millennia by micrometeorite bombardment into soft, gently rolling contours, and occasional large meteorites have blasted out fresh craters, but otherwise the Moon has remained unchanged, silent, spinning eternally through space.

About 300 million years ago, when land animals and plants were becoming established on Earth, an unusually large impact produced the crater Tycho. Half a dozen times in the last decade, animals from Earth landed on the surface in frail, glittering craft, scrabbled around briefly on the surface, and then left. The odd bits of gear that they left behind, and the descent stages of their spacecraft, will remain on the Moon long after man as a species has ceased to exist.

One of the spacecraft left on the Moon carries a plaque. It says, 'We came in peace for all mankind.' But who will read it?

4. Mercury

Mercury is one of the few celestial objects that the ancients had a better chance of observing for themselves than we do today. The planet orbits very close indeed to the Sun (its mean distance is only 58 million kilometres), which means that it is never far from the Sun in the sky, and is almost always lost in its glare. The only occasions when Mercury can be seen at all clearly are when it is at its greatest apparent distance from the Sun, and then only at sunrise and sunset. In good conditions, Mercury can then be seen shining like a tiny pink lamp against the glow of the sky. All too often, of course, it is obscured by clouds, by smoke and fumes rising above our big cities, and by the harsh glare of street lights. How fortunate were those early observers who could see the planet gleaming brightly against an undefiled evening sky!

Because of its characteristic appearance in two separate guises, as an 'evening' star after sunset and a 'morning' star before dawn, Mercury was thought to be two separate bodies by the earliest civilizations. The Greeks, although aware that it was a single entity, retained two names for the planet, calling it *Hermes* as the evening star and *Apollo* as the morning star (Hermes was the winged messenger of Zeus; Apollo was the Sun-god). Heraclitus argued that, because Mercury is always so close to the Sun in the sky, it must revolve around the Sun and not the Earth, a perceptive analysis which was a precursor to later ideas about the sun-centred solar system.

Mercury is a small planet. Its diameter is only 4,880 kilometres, not much bigger than our Moon. Lacking a moon of its own, Mercury's mass was difficult to determine. It was eventually

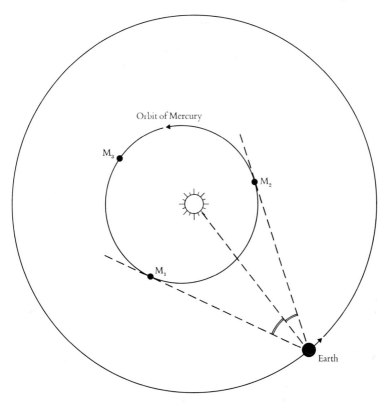

Figure 4.1. *When seen from the Earth, Mercury appears to be at its greatest distance from the Sun when it is at positions M_1 and M_2. At all other points on its orbit, such as M_3, it will appear to be nearer the Sun in the sky, although, in terms of absolute distances, there is little difference between M_1, M_2 and M_3.*

found by measuring the changes in Eros' orbit – the same asteroid that was used to calculate the value of the astronomical unit – caused by the gravitational force acting between Mercury and the asteroid. The mass was found to be only $3 \cdot 3 \times 10^{23}$ kg, making Mercury the second smallest planet in the solar system. (Only Pluto is smaller.)

Mercury takes eighty-eight days to complete one orbit around the Sun. Mercury's orbit is much the most elliptical of all the planets. The long axis of the ellipse drifts slowly around the Sun, at a rate of about 10 minutes of arc per century. This unspectacular and perhaps rather esoteric statistic has been of immense importance in astronomy, and in science as a whole.

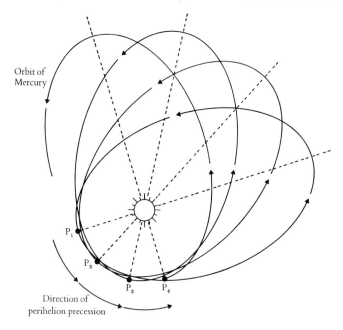

Figure 4.2.   The precession of Mercury's perihelion. Successive perihelion points for each orbit of Mercury around the Sun are shown as P_1, P_2, P_3, etc. Although the orbit is elliptical, the perihelion point drifts around Mercury on a roughly circular path.

In 1845, the French astronomer Urbain Jean Joseph Le Verrier was investigating the motions of planets, and the gravitational disturbances that they cause to each other's orbits. When dealing with Mercury, he found that he could explain almost all of the

10 minute per century drift of its long axis by taking into account the gravitational influence of Venus and the Earth. These together could account for about 9 minutes and 25 seconds of the drift, but 35 seconds remained unaccounted for. Le Verrier concluded that there must be *another* planet between Mercury and the Sun.

Le Verrier deduced correctly that, if such a planet existed, it should occasionally be observed as a tiny black spot drifting in front of the Sun. Such black spots had been observed on many occasions in the past without being satisfactorily explained. Le Verrier sifted through all the records of such observations, many of which were clearly due to poor observation (spots in front of the eyes!), and concluded that there was indeed a small planet between Mercury and the Sun. This was christened *Vulcan*, and Le Verrier confidently predicted that it went round the Sun every thirty-three days, and that it would cross in front of the Sun on 22 March 1877.

Sadly, nothing at all was seen that day, nor has 'Vulcan' ever been observed on any subsequent occasion. It should not be thought, though, that Le Verrier was an eccentric crank in his diligent studies of a non-existent planet; by applying exactly similar computational techniques, he succeeded in discovering Neptune in 1846.

Le Verrier's original thirty-five-second discrepancy in the drift of Mercury's long axis (known formally as the *precession of the perihelion*) remained unexplained after the demise of Vulcan. As studies were refined during the nineteenth century, the size of the discrepancy increased to about forty-two seconds. This anomaly remained unexplained until 1915, when no less an eminence than Albert Einstein appeared on the scene. At that time, he was formulating his General Theory of Relativity, which was to change the whole nature of physics.

To demonstrate the validity of his theory, Einstein made three separate predictions which could be checked by independent observations. He considered first the precession of Mercury's perihelion. He calculated from his theory that the rate of advance should be around 43 seconds of arc per century; amazingly close to the *observed* value.

One of Einstein's other two predictions based on relativity theory was that a ray of light passing near a massive body such as the Sun would be deflected by its enormous gravitational field. This was demonstrated initially by observing the rays of light from stars near the Sun during a total eclipse. Small apparent shifts in the stars' positions were observed, corresponding to the bending of the light rays from them by the Sun (Figure 4.3). Unfortunately, the shifts were so small that some critics were able to argue that they were caused by observational errors. More recently, Mercury has been used in a further test of this aspect of relativity theory. Radar signals were sent out from Earth to

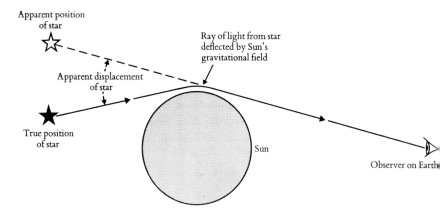

Figure 4.3. One of Einstein's tests of relativity theory. He predicted that light from a star would be deflected by the enormous gravitational field of the Sun, so that it would appear to be coming from a slightly different position. The apparent position of the star would appear to be shifted relative to other stars whose light does not pass near the Sun.

Mercury when the planet was very nearly in line with the Sun, and behind it. Since the size and shape of Mercury's orbit is now known precisely, it was possible to predict precisely when a radar echo from Mercury should be obtained. In the experiments, a distinct lag was observed, corresponding to the extra distance travelled by the radar signals in being deflected towards the Sun. Einstein was right again.

The rotation period of Mercury: a sad scientific saga

The present generation of studies of the solar system has shown up a number of misconceptions and outright errors that grew up in the early years of observational astronomy, and subsequently became entrenched in the literature. One of the most fascinating cases concerned the rotation period of Mercury.

It was remarked at the beginning of this chapter that Mercury is a difficult object to see with the naked eye. With a powerful telescope, observing Mercury is even more difficult, for, when it is low in the sky, its image boils and swirls in the turbulence present along the long optical line of sight through the Earth's atmosphere. Armed with a calculator to work out the right coordinates, it is possible to find Mercury in the sky in broad daylight, and some useful telescopic observations have been made in these conditions, because when the planet is high in the sky there is much less atmosphere present along the line of sight, and therefore less turbulence.

Even in the best possible viewing conditions, though, it is difficult to make out surface details on Mercury. Fuzzy drawings showing some features were first published at the beginning of the nineteenth century, and these were used by a German astronomer, Friedrich Bessel, to derive a rotation period of exactly twenty-four hours. Since the surface features were only very indistinctly visible, it was extraordinarily difficult for Bessel to judge when the same feature had rotated round into the same relative position, and he was almost certainly influenced in his work by the knowledge that both the Earth's and Mars' rotation periods are also twenty-four hours.

Bessel's result was accepted until 1891, when the Italian astronomer, Giovanni Schiaparelli (who is better known for his classic studies of Mars), made a fresh series of observations and decided that the rotation period could not possibly be as short as twenty-four hours. Instead, he proposed that it was eighty-eight *days*, exactly the same as the length of the Mercurian year. In suggesting this, Schiaparelli may have been influenced by work going on at the time into the reasons for the spin–orbit coupling shown by the Moon, which, of course, keeps one face permanently turned towards the Earth. Schiaparelli decided that Mercury must also have one face turned permanently towards the Sun. An obvious implication of this was that one side of Mercury should be blisteringly hot, while the other side, away from the Sun, should be freezing cold.

Amazingly, Schiaparelli's work was not seriously questioned until 1965. In 1962, American scientists at the University of Michigan had commenced studies of radio emissions from Mercury to determine the surface temperature of the planet. Naturally, they expected to find one 'hot' side and one 'cold' side. Much to their surprise, their measurements showed an average temperature of around 130 °C, with no major difference between the 'hot' and 'cold' sides of the planet. This should have given astronomers pause for thought, but so committed was the astronomical community generally to the idea that Mercury always kept one face towards the Sun, that an alternative explanation was cooked up to account for the inconvenient observations: Mercury, it was declared, had an atmosphere which was capable of circulating from the sunlit side of the planet to the dark side, carrying heat with it, and thus equalizing the temperature.

This idea was not scotched until American radar workers, using the same techniques as those that had been applied to the Moon, began to look at the differences in frequency of radar pulses beamed at the edges of the planet. They found in 1965 that their results were not consistent with an 88-day rotation period, and that the data could only be fitted to a 58·6-day period. This, of course, means that Mercury does *not* always keep one face turned towards the Sun, and therefore the rather improbable idea of a circulating atmosphere could be dispensed with.

On the face of it, 58·6 is rather an odd number. It was not long before Giuseppe Colombo, an Italian astronomer, twigged to the fact that 58·6 is almost exactly two thirds of 88. This could be no coincidence. It means that, in the time that Mercury takes to complete *two* orbits of the Sun (176 days), it rotates on its axis *three* times. This is a slightly more complex example of the phenomenon of spin-orbit coupling than we encountered with the Moon. Like the Moon, the coupling of Mercury's orbital and rotation periods almost certainly arises from tidal forces, acting between Mercury and the Sun. Operating over a long period, the tidal forces slowly braked Mercury's spin until a least-energy condition was established in the resonance between spin and orbit.

The fact that Mercury's day lasts two thirds of its year has some surprising implications. Figure 4.4 shows how the Sun *appears*

Figure 4.4.   The Sun as seen by an observer on Mercury throughout two Mercurian years. Initially, the Sun would seem large and near; subsequently, it would shrink and appear more distant.

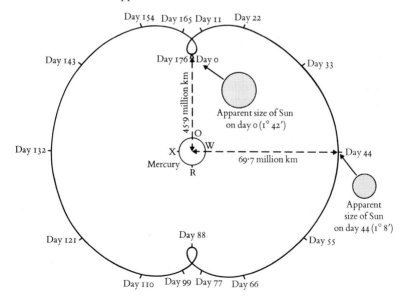

to move relative to Mercury throughout two Mercurian years. On two occasions, eighty-eight days apart, the Sun appears to come to a halt in the Mercurian sky, to backtrack slightly and then continue. The distance between Sun and Mercury also varies. To an observer living on Mercury at the point labelled O, the Sun would appear to rise over the horizon as a small, bright body, swelling steadily in size as it approached the zenith, and slowing down as it did so. At about 1 degree past its zenith, it would stop in its tracks, slide backwards for a degree, stop again, and then move forward in the original direction, accelerating and getting smaller as it did so.

An observer living at W would see a completely different and equally bizarre sequence of events. The Sun would first rise, looking very large, show its face over the horizon briefly, then duck down below it again for a short while, and finally rise again, drifting slowly across the sky and shrinking in size until it reached the zenith. When it passed the zenith, the same sequence of events would be reversed, with a false sunset taking place before it finally disappeared, not to rise again for a whole year.

Clearly, keeping a calendar on Mercury would be a complicated and confusing affair. More important than this, however, are the thermal effects that would arise. The areas on Mercury immediately beneath the positions where the Sun does its little somersault (O and R on Figure 4.4) clearly receive more sunshine hours than those at W and X, and, furthermore, the Sun is also much closer when it is opposite O and R. Hence the amounts of solar radiation falling on O and R are much greater – in fact, about twice as great – than those falling on W and X.

O and R and W and X are, in three dimensions, lines of longitude on the surface of the planet. The points of intersection of O and R on the equator of Mercury mark the places where the *maximum possible* solar radiation falls. They have very aptly been described as 'hot poles'. It is important to grasp that they are poles, actual physical points on the surface of the planets, and not abstract, moving points. Mercury's spin–orbit coupling ensures that the hot poles are locked into the same places relative to the Sun permanently. Every time the Sun comes nearest, and appears to come to rest in the sky, it is the hot poles that its rays beat against. The temperature at the hot poles is believed to rise as high as 430 °C, more than enough to melt lead.

W and X by contrast represent areas of minimum temperatures, and are usually called 'warm poles'. During the Mercurian night, the temperature there may fall as low as −173 °C. On Earth, we are familiar with a regular procession of seasons. These are a consequence of the Earth's axial inclination – 23½ degrees – which is also responsible for the variation of temperature with *latitude*. Seasons also exist on Mercury (extremely hot, hot and less hot), but these vary with both latitude and *longitude*. Thus a hypothetical Mercurian going on vacation to do some sunbathing would not simply head towards the equator, as we do on Earth, but would head either east or west, aiming for the nearest 'hot' pole (Figure 4.5).

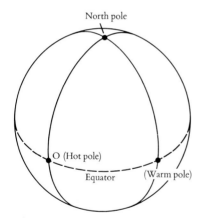

Figure 4.5. The hot and warm poles on Mercury. At the hot pole, the temperature may vary during a Mercurian day by as much as 600 °C. This huge variation must have considerable effects on the surface materials.

Mercury before Mariner 10

Mercury was rather an unknown quantity before the advent of space missions, as the confusion over its orbital period reveals. It was impossible to say much about the surface features, even when the rotation period was accurately known, because even the largest telescopes could not show features less than 300 kilometres across. There was some evidence, however, that Mercury was similar to the Moon. Photometric studies showed that, like the Moon, it reflects only 7 per cent of the sunlight falling on it, and polarimetric observations indicated a churned-up, dusty surface like the lunar regolith.

The density of Mercury was easily determined once the mass had been found. At about 5.4×10^3 kg m^{-3}, it is close to the Earth's, and suggests the presence of a large metallic core with a relatively thin silicate crust and mantle. The confusion over the orbital period of Mercury led to some confusion over the presence or absence of an atmosphere. In the days before radio emission studies were made, it was supposed that one side of the planet was extremely cold, so, despite the relatively small mass, it was considered that an atmosphere might be present, equivalent to about 10 millibars of carbon dioxide. There were even some claims that the surface features had been seen to be obscured by clouds or 'veils' of dust.

More rigorous observations, however, particularly of transits of Mercury across the Sun, failed to show the slightest signs of any refraction effects, and the radar data eventually demonstrated that Mercury lacked an atmosphere. Very little else could be deduced. It was not possible to say anything about the composition of the surface rocks, or their age, or whether the surface was cratered or not, or if the planet possessed a magnetic field. Mercury, therefore, was an unknown planet, ripe for exploration.

Mariner 10: 'an exquisite celestial slingshot'

If one were asked to name the single most significant space mission in the brief history of space science, it would be difficult to decide whether this accolade should be given to Mariner 9 or Mariner 10. Both of them were hugely successful missions, and both of them provided stunning new perspectives of alien worlds. Mariner 9 provided a greater range of information than Mariner

10, but Mariner 10's achievement was perhaps the greater, because what it showed was *totally* new – it is easier to observe Mars with a telescope than Mercury, and earlier Mariner missions had provided some brief glimpses of Mars' surface.

Mariner 10 was a beautifully conceived mission. It was launched in November 1973, and its trajectory was calculated so that, three months later, it passed within 6,000 kilometres of Venus, obtaining some superb pictures. In sweeping through space so near to Venus, the spacecraft's velocity and direction were changed

Figure 4.6. Mariner 10's trajectory was so exquisitely calculated that, after swinging past Venus, it flew past Mercury on three separate occasions: 29 March 1974, 21 September 1974 and 16 March 1975.

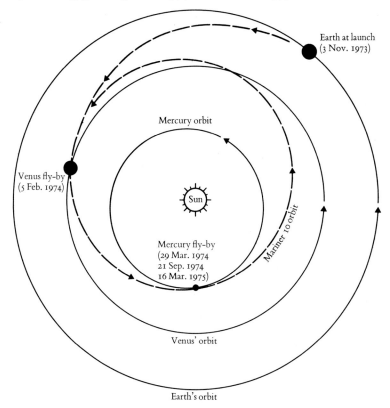

Earth at launch
(3 Nov. 1973)

Mercury orbit

Venus fly-by
(5 Feb. 1974)

Sun

Mariner 10 orbit

Mercury fly-by
(29 Mar. 1974
21 Sep. 1974
16 Mar. 1975)

Venus' orbit

Earth's orbit

and it was flung off like a slingshot into a new trajectory which eventually carried it into orbit around the Sun. This orbit was carefully chosen so that the spacecraft came into the vicinity of Mercury every 176 days, or once every two Mercurian years, and thus it was able to 'encounter' Mercury on no less than three occasions before it ceased operating.

The spacecraft carried a number of different instruments: TV cameras to scan the surface, a spectrometer to seek any traces of atmospheric gases, an infra-red sensor to examine surface temperature and a magnetometer to check on magnetic fields. Together the results of all these observations combined to produce an enormous increase in our knowledge of Mercury. No longer was it an obscure little planet, lost in the glare of the Sun. Suddenly, it became a 'new' planet, with an identity all of its own.

The new view of Mercury

The scientists of the Mariner 10 team who crowded into the mission control room at the Jet Propulsion Laboratory, Pasadena, on the occasion of the first encounter in March 1974, had a unique privilege. To them fell the extraordinary opportunity of watching a stream of pictures come in from the spacecraft as it approached Mercury; of seeing the planet swell from a small white spot to fill the frame of the camera; and finally of gazing on the surface of the planet for the first time. The sense of awe and excitement that filled the room can well be imagined. It was a moment unrepeatable in history.

On its first encounter, the spacecraft swept past the equatorial regions of the planet at a distance of about 700 kilometres; on its second encounter it passed over the south pole at a distance of 50,000 kilometres; but on its third, it grazed the northern hemisphere at a mere 350 kilometres. The two low passes enabled detailed pictures of the surface to be obtained. Several thousand good-quality pictures were obtained, sufficient to keep teams of photo-geologists busy for years. This was in itself something of a minor embarrassment, since, with a flood of lunar pictures still not fully assessed and a fresh deluge of Mariner 9 pictures still to be processed, life was more than a little hectic for the teams involved. By the end of the third encounter, photographic coverage of about 40 per cent of the planet's surface had been obtained.

The pictures revealed a barren, desolate world, a harsh outpost of the solar system, scorched by its proximity to the Sun. Superficially, Mercury looks very like the Moon – a barren sphere pockmarked with craters. There are even examples of rayed craters, splashing white streaks over the surface. At first sight it is often difficult to distinguish between close-up pictures of Mercury and the Moon. A second, closer inspection reveals some important differences.

First, in the chapters on the Moon, it was emphasized that two major terrain types exist there: the pale-coloured cratered highlands, and the dark-coloured lava plains of the maria or 'seas'. No such obvious contrast exists on Mercury. There are no comparable smooth, dark expanses of lavas – and hence no 'man in Mercury'.

Secondly, while there is unambiguous evidence of relatively young volcanic activity on the Moon (forming the lavas of the maria), the situation on Mercury is a good deal more complex and is not yet fully understood.

Thirdly – and predictably – there are some subtle differences between craters on the Moon and on Mercury. These are caused by the differences in surface gravity. Since Mercury is denser and more massive than the Moon, its surface gravity is two and a half times higher, and material flung out of impact craters therefore travels less far than it would on the Moon.

The Mariner 10 pictures provided as good a view of the battered surface of Mercury as can be obtained of the Moon with a high-powered telescope. Anyone who has looked at the Moon with even a small telescope, and seen the wealth of detail revealed there, will have a good idea of the scale of the task facing the Mariner 10 mission scientists. Not only did they have to map and name all the hundreds of surface features, effectively establishing

Figure 4.7.   A photomosaic of one hemisphere of Mercury, built up from pictures taken by Mariner 10 as it moved away from the planet after its first encounter. The similarities with the Moon are obvious.

the *geography* of the planet, but they also had to try and work out something of its *geology*.

In mapping any planet, the first step required is to establish a set of lines of latitude and longitude. This was done several centuries ago on Earth, and the line of zero longitude, or prime meridian, was eventually decided by using the Royal Observatory at Greenwich as a fixed reference point. (Any other point would have done equally well – Paris was preferred for many years.) On Mercury, a small but fairly distinctive crater was selected as the fixed reference point to define the zero meridian in the planet's set of lines of longitude. The crater is called Hun Kal.

Hun Kal is not the sort of name that immediately comes to mind. Who on Earth – or on Mercury, for that matter – was Hun Kal, one might ask? Or, even, *what* was Hun Kal? This raises one of the less profound and more entertaining problems created by the modern era of spacecraft exploration. How do you set about giving names to all the features on a newly discovered body? The task is very great. Imagine having to think of names for all the mountains, rivers, hills and valleys in Europe, just for a start! How would you set about it? Well, of course, you would set up a nomenclature committee. Inevitably a committee. To be fair, the Mercury Nomenclature Committee, which works under the auspices of the International Astronomical Union, has made a good start and has introduced some rather original ground rules.

The main valleys are being named after famous Earth-based radio astronomy observatories, such as Arecibo, Goldstone and Crimea, while scarps are named after ships which took part in the great voyages of discovery on Earth: Fram, Endeavour, Vostock, Santa Maria, and so on. For plains, it has been decided to use the names of the planet Mercury in different languages. Tir, Bouda, Odin and Sobkolu have already been christened. Sadly, no rigorous scheme for naming craters seems to have emerged. No one seems to know who or what Hun Kal is or was . . .*

---

* Exhaustive researches have revealed that Hun Kal is 'twenty' in the extinct language of the ancient Maya civilization. The zero meridian is set 20° from the crater.

Craters and the heavily cratered terrain

Large parts of Mercury are thickly covered with craters of all sizes, from tiny pits to enormous basins hundreds of kilometres across. They closely resemble parts of the highland areas of the Moon (Figure 4.8). Like the craters on the Moon, the largest Mercurian craters have much more complex structures than the smallest – they have central peaks, produced by the reflection of shock waves during the impact process, and their walls have several concentric terraces. Like the Moon, too, the youngest Mercurian craters are crisply defined with ray systems radiating away from them. Older ones have less prominent rays, and as they become older, take on a progressively more battered, weathered appearance. The 'weathering', though, is purely the result of impact processes and is brought about by aeons of micrometeorite bombardment. There is no evidence whatever

Figure 4.8.   Part of Mercury's heavily cratered terrain. Although there is a strong superficial resemblance to the lunar highlands, there are important differences. It may be significant that the smaller craters are better formed than the older ones. The old, large craters have poorly defined rims and lack central peaks. This suggests that the early crust of Mercury was rather plastic, and that Mercury may not have quite such an ancient surface as the Moon. The surface is, however, still more than 4 billion years old.

for any other kind of erosion having taken place on Mercury – if the planet ever had an atmosphere, it must have lost it very early indeed in its history.

To detect differences between lunar and Mercurian craters requires sharp eyes and systematic measurements. The *secondary* craters on Mercury, caused by ejected material falling back around primary impact sites, are much less widespread than on the Moon. Mercurian craters of any given diameter also tend to be shallower than their counterparts on the Moon. Both these differences are attributable to the higher surface gravity on Mercury than on the Moon.

The Caloris basin

The most spectacular features on the Moon are, without doubt, the large circular ring basins, such as the Orientale Basin (on the far side) and the Mare Imbrium. Mercury possesses only one comparable structure, the Caloris (or 'Hot') Basin. It is some 1,300 kilometres in diameter, and is surrounded by a single ring of high mountains rising some 2 kilometres above the plains inside the basin. The mountains are thought to be blocks of the Mercurian crust heaved up by the impact event that excavated the Caloris Basin. This catastrophic impact blasted out huge quantities of ejected material, much of which fell back around the impact site to excavate conspicuous ridges and grooves radiating around the basin. This terrain, which is analogous to that around the Imbrium Basin on the Moon, has been called 'lineated terrain'.

The impact which created the Caloris Basin was so enormous that its effects are visible on the opposite side of the planet. In the antipodean region of the basin, there is an area of chaotic or 'weird' terrain, which is believed to have been thrown up by shock waves which travelled through the centre of Mercury and were focused on the point of the surface immediately opposite the impact site. It is as if an impact taking place in Britain had caused a jumble of small blocks and hills to be thrown up in New Zealand.

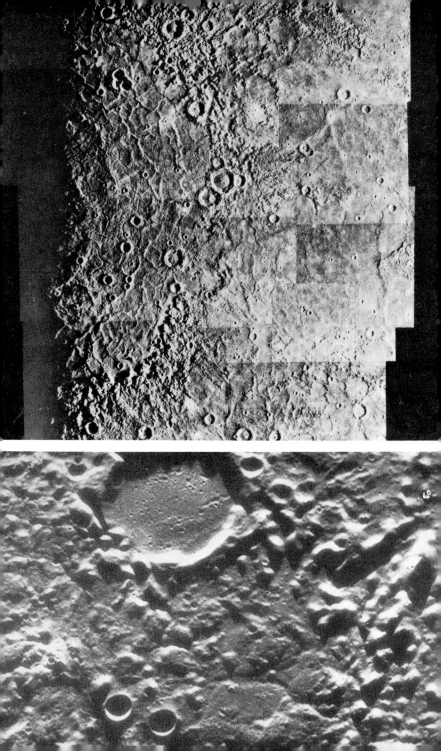

Figure 4.9. (opposite)   *The gigantic Caloris Basin, of which only about one third is visible. Note the radial lineated terrain, and the areas of polygonal fractures.*

Figure 4.10. (opposite below)   *This jumble of hillocks and hollows lies on the far side of Mercury, directly opposite the Caloris Basin, and constitutes part of the so-called 'weird terrain'.*

Smooth plains

The Imbrium Basin on the Moon is entirely filled with basaltic lavas which admirably justify the description 'smooth plains'. Such plains also occur within and surrounding the Caloris Basin, but some Mariner 10 scientists have had reservations about the presence of lavas here. They *look* rather similar – rather featureless expanses peppered with small impact craters – but there is no sign of obvious volcanic features such as the lava-flow fronts on the Mare Imbrium. Another worrying feature is that, on the Moon, the volcanic rocks of the maria are definitely darker than those of the non-volcanic highlands. No such obvious contrast exists on Mercury.

Figure 4.11.   Part of the smooth plains. Although there are some similarities with the lava-covered lunar maria, it is uncertain whether the Mercurian smooth plains are of volcanic origin.

The Mariner scientists are therefore treading cautiously in their interpretation of the smooth plains. Some feel that the similarities with the Moon are close enough to indicate that the smooth plains are also volcanic. Others, however, remembering the salutary lessons of Apollo 16, where some rocks in the lunar highlands initially thought to be volcanic were subsequently proved to be impact breccias and melts, think that the Mercurian smooth plains might also be impact breccias and melts, and not volcanic. Smooth plains, however, cover vast expanses on Mercury, whereas the problematical Cayley Plains of Apollo 16 were comparatively small and insignificant. Hence, the impactists may be rather over-reacting to the Apollo 16 upset. The problem is an important one, but it will probably not be fully resolved until spacecraft have landed on Mercury and sampled the surface.

The inter-crater plains

In many ways these are even more difficult to interpret than the smooth plains, but may be even more important in understanding the history of Mercury. A glance at any picture of the lunar highlands shows that they consist of nothing but large craters. Craters abut against one another, cut across one another and are even superimposed upon one another. There are few areas which could be said to be 'between craters'. This abundance of large craters is interpreted as being the result of *saturation*; so many impact events took place to form the lunar highlands that a point was reached where every new impact created a fresh crater only by obliterating earlier ones.

Although the heavily cratered areas on Mercury closely resemble the lunar highlands, there are large parts of the surface where there appear to be 'gaps' between the craters, and it is these regions that are known as *inter-crater* plains. Such plains are in fact cratered – they are peppered with *small* craters, but they represent spaces between large craters which have not been 'filled in' with other large craters.

Much debate surrounds the origin of the inter-crater plains, and it involves some fundamental questions. Was there only one

Figure 4.12.   An example of the so-called 'inter-crater plains'. Although this is clearly an ancient surface, there are areas between the large craters which are scarcely cratered at all, a situation which is not found in the lunar highlands.

Figure 4.13.   Another view of the inter-crater plains. This picture illustrates the vast areas of Mercury covered by terrain of this type.

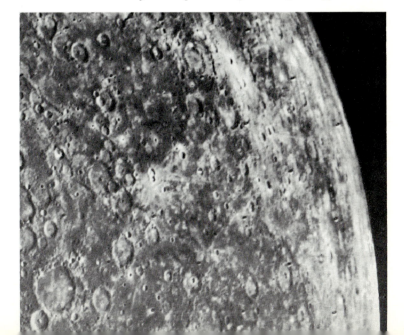

episode of major impact events? Or were there two or more separated by intervals of inactivity? What happened to the *rate* at which bombardment took place? When did it start to decay? There are no easy answers to these questions, but two separate hypotheses have been advanced to account for the inter-crater plains.

The first argues that the early Mercurian surface was never saturated with craters, that the inter-crater plains therefore represent an extremely early surface of the planet, which has not been nearly as extensively modified as the lunar highlands. If this is the case, then, in the Mercurian inter-crater plains, we have a glimpse of what must be one of the oldest surfaces in the solar system, a look back through 4,600 million years of history.

The alternative hypothesis is quite different. According to this, the first surface of Mercury *was* obliterated – or saturated – with impact craters, but subsequently some kind of event took place to smooth out the first formed craters. This could have taken the form of a planet-wide episode of volcanism. Subsequent to this smoothing out, further bombardment, much less intense, and not even approaching saturation, excavated the large craters that we see at present. Again, it will require surface sampling missions to Mercury to resolve which – if either – of these ideas is right.

Scarps

The scarps of Mercury are not immediately eye-catching features. They are long, low breaks or steps in the topography, which can be traced for hundreds of kilometres. Nothing quite like them exists on the Moon, Mars or even the Earth. It is thought that these scarps were produced early in the history of Mercury, perhaps when it had an entirely liquid interior. As it cooled, the planet shrank and decreased in volume. This caused the thin, brittle outer crust to wrinkle up into ridges and scarps, rather in the way that the skin of an apple wrinkles and pleats as the inside dries out if it is kept in the fruit bowl too long.

Figure 4.14. A good example of a Mercurian scarp is seen here, running diagonally towards the skyline. The cliff line is only a few hundred metres high, but extends for hundreds of kilometres.

Mariner 10: other instruments

Apart from the cameras which transmitted to Earth such a wealth of new data, Mariner 10 carried several other important instruments. By far the most important discovery to come from these was that Mercury has a magnetic field. This was completely unexpected, because theoretical considerations had suggested that the planet ought not to have one.

Mercury's magnetic field is not very strong, to be sure – it is only about one hundredth of the intensity of the Earth's – but it does have north and south poles, which, like the Earth's, are closely aligned with the planet's rotation axis. Now the Earth's magnetic field, it is supposed, is generated in the core by slow-moving currents of molten metal, whose motions are in part induced by the Earth's rotation. The density of Mercury suggests that it, too, has a large metal core, which may also be liquid. But because the planet rotates so slowly (fifty-nine times slower than the Earth), it had been thought prior to Mariner 10 that there could not be any motions in the core capable of generating a field.

The discovery that a significant field did in fact exist created a considerable stir, and scientists were forced to do a lot of quick rethinking. As with the Moon (whose field is *much* weaker than Mercury's), the main alternative ideas are that Mercury's field is a 'fossil' one, acquired early in its history and never subsequently lost, or that the field has been induced somehow by interactions between the planet and the Sun's magnetic field. Neither of these explanations is particularly satisfactory. But, as is usually the case when preconceived ideas are upset, the discovery of the existence of Mercury's magnetic field has provided an excellent spur to further investigation. As a result, we will almost certainly learn more about the nature of planetary magnetism in general, and perhaps even about the origin of the Earth's magnetic field.

Although there had been some rather wild hypothesizing about the existence of an atmosphere of sorts on Mercury when the planet's rotation period was still in doubt, there were good grounds for thinking that Mercury effectively lacks an atmosphere. Mariner 10 carried an ultraviolet spectrometer to look for traces of an atmosphere, but this revealed no surprises. The atmospheric pressure recorded was only 10^{-12} millibars; 1,000 billionth of the Earth's. The most abundant gas in this exceedingly tenuous atmosphere is helium, which may be trapped in the vicinity of Mercury by the magnetic field, after having been blown out from the Sun in the solar wind.

The evolution of Mercury

There are no samples from the surface of Mercury, nor are there likely to be for many years to come. Our knowledge of the planet is therefore scanty compared with that of the Moon. We can, however, go a long way towards working out the history of

Mercury, and in this our knowledge of the Moon provides both a valuable guide and a standard of comparison.

The first event, of course, was the initial accretion of the planet from the dust of the primitive solar nebula 4·6 billion years ago. It follows from Mercury's proximity to the Sun that, like the Moon, it is likely to be rich in refractory elements, and poor in volatile ones. From considerations of density, it is likely that Mercury has a large heavy iron core surrounded by a thin silicate

Figure 4.15.   A comparison of the internal structure of the Moon and Mercury. Mercury has by far the largest core of any planet, relative to its overall size.

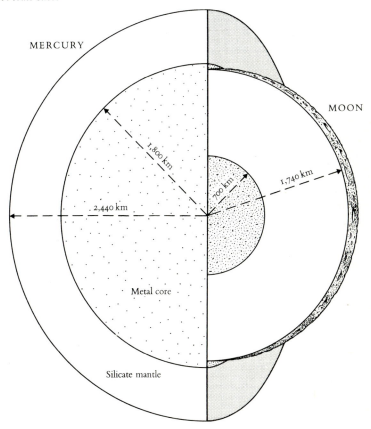

MERCURY

MOON

1,800 km

700 km

1,740 km

2,440 km

Metal core

Silicate mantle

mantle; these must have segregated almost immediately after the accretion process, which may itself have led to planet-wide melting. A thin crust of igneous rocks formed on the outer surface, to be immediately bombarded by in-falling asteroidal materials. If Mercury ever had an atmosphere, it must have been driven off during these early days in its history.

What happened next depends on one's interpretation of the inter-crater plains. Either there was some kind of episode which blotted out the first-formed craters, or the intensity of bombardment was never great and the surface never became saturated with large craters. By analogy with the Moon, the early cratering which gave rise to the heavily cratered terrain probably took place between 4·6 and 4·0 billion years ago.

At the end of this cratering episode, the giant Caloris Basin was formed, probably at roughly the same time as the Imbrium Basin was formed on the Moon. This sculpted the huge area around the Caloris Basin, excavated the 'lineated terrain', and threw up the weird terrain in the antipodean regions.

At the same time, or shortly afterwards, the 'smooth plains' were formed, possibly by volcanic activity, possibly not. It is difficult to put a precise age to these surfaces. Crater-density studies suggest that they are slightly older than the lunar maria, but there is no guarantee that the bombardment rate was uniform throughout the solar system.

Finally, during the 4 billion years after the smooth plains were formed, nothing much happened on Mercury, save for a light sprinkling of impact craters. The youngest of these, those with bright ray systems, are probably comparable in age with the young rayed craters on the Moon – say, a few hundred million years. Mercury, it seems, is a pretty dead planet.

5. Venus

Most of the planets are difficult to see: Venus is hard to avoid. It comes so close to the Earth that it often hangs as a brilliant jewel in the evening or morning sky, blazing brightly enough at night to cast its own shadows, and to create a slim shivering track on water. Its beauty is so outstanding that, through the centuries, it has driven poets to compose extravagant romantic verses about it; the planet seems to have been associated with the goddess of love since at least 3000 B.C. Like Mercury, the Greeks originally had two names for Venus. As an evening star, they called it *Hesperus*, and as a morning star *Phosphorus*. The Romans called it *Lucifer*, the bringer of light. A more modern consequence of Venus' brilliance is that it is regularly reported as an unidentified flying object (UFO), notwithstanding the fact that it hangs motionless in the sky, and can be seen night after night in much the same place.

Venus orbits at a mean distance of 108 million kilometres from the Sun. Consequently, at the times of its closest approach, it comes within 40 million miles of Earth. Its diameter is 12,100 km, its mass 4.9×10^{24} kg★ and its density about 5.2×10^3 kg m^{-3}. All these statistics add up to make a planet that is so similar to the Earth that it is tempting to think that Venus might be a world like our own, with continents and oceans, forests and streams, flowers and plants.

Seen through a telescope, Venus does little to dispel this hope. It shines serene and immaculate, its external beauty encouraging one to believe that it is, indeed, a wholly agreeable companion to

★ The mass of Venus was first estimated in 1757 by Alexis Clairaut, who measured the perturbations of Venus' orbit caused by the Earth. He concluded that Venus' mass was two thirds that of the Earth.

the Earth. Even the earliest observers, however, realized that Venus' flawless appearance was created by the fact that they were observing only the surface of a thick layer of clouds rather than the solid surface of the planet. Christian Huygens wrote in the seventeenth century: 'I have often wonder'd that when I have view'd Venus . . . she always appeared to me to be all over equally lucid, that I can't say I observed so much as one Spot in her . . . is not all the Light we see reflected from an Atmosphere surrounding Venus?'

The presence of such a thick, cloud-laden atmosphere naturally encouraged early astronomers to believe that Venus was indeed a congenial place, perhaps wetter and more humid than the Earth, but still basically a pleasant, Earth-like place.

Nothing could be further from the truth. Spacecraft surveys have revealed that Venus, far from being an idyllic abode for lovers is, as one distinguished American scientist put it, 'astonishingly hot, with an oppressively dense atmosphere containing corrosive gases, with a surface glowing dimly by its own red heat and characterized by bizarre optical refraction effects. Venus . . . seems very much like the classical view of hell.'

The early days

Our view of Venus has changed dramatically as a result of space missions, but it is instructive to look back at how this view evolved. Because Venus is so bright – it reflects nearly 80 per cent of sunlight falling on it, compared with the Moon's 7 per cent – it was naturally one of the first objects on which astronomers turned their telescopes.

Galileo – who else – was first off the mark, and made the simple but important observation that Venus, like the Moon, goes through a series of phases, changing from a slim crescent to a full disc in 292 days. As Galileo rather cagily put it: *Cynthiae figuras aemulater mater amorum* ('The Mother of Love imitates the form of Cynthia', that is the phases of the Moon). Galileo wrapped up his simple observation in tortuous language because he lived in sensitive times, when his astronomical work was bringing him into a headlong clash with the established Church.

The observation that Venus exhibits phases is of much more profound significance than it may at first seem. Those who believed that the planets revolved around the Earth, not the Sun, were reinforced in their belief by the fact that Venus always looks almost equally bright. If it revolved around the Sun, and not the Earth, they argued, then its brightness should vary as its distance from the Earth varies. And it does not.

Figure 5.1. The phases of Venus. When close to the Earth, Venus appears only as a thin crescent; when on the opposite side of the Sun, it appears as 'full'. Hence, its apparent brightness does not change as much as early astronomers argued that it should, if it orbited round the Sun and not the Earth.

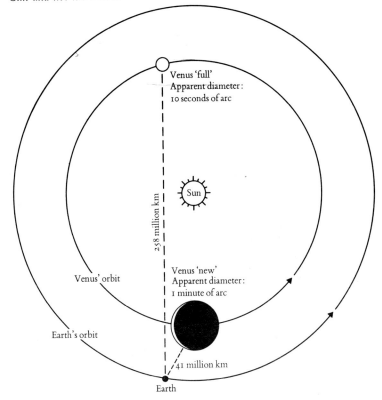

What Galileo discovered was that when Venus is nearest the Earth, it exhibits a slim crescent; when it is furthest away, on the opposite side of the Sun, it presents its full face to the Earth. So, although the distance between Earth and Venus increases vastly between crescent and full, the *brightness* remains about the same because more of Venus is lit up the further away it is.

After Galileo, it was difficult for observers to find out much more about Venus. This is not to say that they did not 'see' much more. As early as 1727, an Italian astronomer had produced a map of Venus which showed 'continents' and 'oceans'. What he was mapping were almost certainly optical defects produced by his telescope. Much later, Percival Lowell, better known for his ideas on the canals of Mars, claimed that he could see thin linear features and circular patches on Venus also. At the time of his death in 1916, he remained convinced of the genuineness of his observations. Later still, other astronomers thought that they could make out permanent features on the surface, which, they postulated, might be true surface features dimly seen through the cloud veils. None of these observations is now taken seriously. At best, the features observed were merely atmospheric disturbances.

A further instance of astronomers seeing what they wanted to see took place in the eighteenth century. Venus was known to have strong similarities with the Earth. Because the Earth has a splendid Moon, it was natural to suppose that Venus might also have one. Careful searches were made for such a satellite, and were successful – apparently. A satellite was described by reputable astronomers such as Giovanni Cassini (famous for his work on Saturn's rings), and in 1761 its existence was regarded as unquestionable when a French astronomer described observations of it over a period of a month. Whatever it was that these astronomers were seeing – and they were honest, hard-working observers – the satellite was never subsequently detected. It seems likely that they mistakenly identified faint stars as the satellite, or possibly 'ghost' images of Venus produced by the poor lenses of their telescopes.

The 1760s was a critical time for Venus watchers. In 1761, and again in 1769, two very rare events took place: Venus passed in

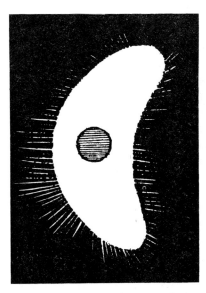
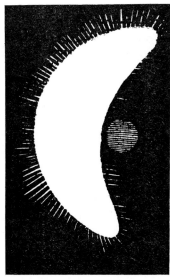

Figure 5.2.   Two early views of Venus drawn by Fontana in 1645. The crescent shape is accurately observed, but the circular splodges are spurious, probably the result of defects in the telescope optics.

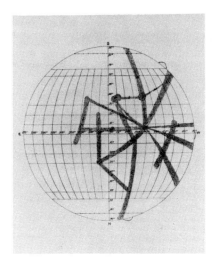

Figure 5.3.   An unusual view of Venus. This telescopic study of Venus was drawn by Percival Lowell in 1896. It bears a striking similarity to his maps of Mars, which are cobwebbed with thin streaks or 'canals' (see Figures 6.9 and 6.10).

front of the Sun. These transits take place at intervals of more than a century, but when they occur, they always do so in pairs. The transits of the 1760s offered major opportunities to make observations of Venus' orbital details, and to refine measurements of the distance between Sun and Earth, so special efforts were made to observe them. Unfortunately, the transits could only be seen from restricted parts of the Earth – rather like the paths of total solar eclipses – so special expeditions were mounted in both 1761 and 1769 to ensure good coverage of the events. The leader of the 1769 expedition was none other than Captain James Cook, who sailed in the *Endeavour* to establish an observation point at Venus Point on the island of Tahiti. Venus Point remains to this day as a reminder of one of the first truly scientific expeditions in history.

Although Cook later in the voyage went on to chart New Zealand and visit Australia, the astronomical observations were not so successful, because Venus' thick atmosphere produced refractive effects which made it impossible to determine the precise instants at which the transit began and ended. As Cook meticulously noted in his log: 'We very distinctly saw an Atmosphere or dusky shade around the body of the planet which very much disturbed the time of contacts.' Since Cook's voyage, there have been only two other transits, in 1874 and 1884. The next will not be until the year 2004. The nineteenth-century transits confirmed the thickness of Venus' atmosphere which Cook had so accurately described.

Speculations on conditions on Venus

Venus was a frustrating object for astronomers to observe. Although excellent telescopes were available by the end of the nineteenth century, there was almost nothing new that an astronomer could glean about the planet by squinting at it visually. All he could see was the bland, blank surface of the atmosphere. A lack of hard data, of course, always encourages scientists to polish up their own pet hypotheses, and to build pyramids of speculation on the flimsiest foundation of facts.

Thus, in 1918, the distinguished Swedish chemist, Svante Arrhenius – a Nobel prizewinner – announced that the whole surface of Venus was dripping wet; that a large part of the surface was covered with swamps, and that lowly forms of vegetable life were likely to exist there. He also argued that conditions were likely to be identical all over the planet, and that Venusian life-forms would therefore be stereotyped, since the lack of diverse environments would not have produced the evolutionary pressures that have prevailed on Earth.

By contrast, in the 1920s, two American astronomers, Seth Nicholson and Charles St John, were arguing exactly the opposite: that the surface of Venus was a barren, dry desert, whose surface was whipped by winds which raised clouds of dust into the atmosphere. These two opposing ideas continued to circulate for decades, and the controversy was not really finally resolved until spacecraft missions provided firm data.

The controversy was rooted in a simple problem: of what is the atmosphere of Venus made? If one could *show* that water exists in Venus' atmosphere, then this would be strong evidence for a 'wet' Venus. Conversely, if one could demonstrate its absence, then a 'dry' Venus would be more probable. Now, it is possible in theory to search for water on Venus by telescope. All one has to do is to use a spectrometer to examine the light coming from the planet. Any water vapour in the atmosphere of Venus would absorb some of the light, and this would be apparent in the form of dark lines in the spectrum. The strength of these lines would also be a guide to the amount of water present.

So far, so good. But – and this is the rub – the light from Venus has to pass through the whole thickness of the Earth's atmosphere before it can reach any ordinary telescope, and the Earth's atmosphere itself contains huge volumes of water. Thus any absorption lines in Venus' spectrum could as easily be the result of terrestrial water as that on Venus.

This problem confounded astronomers for years. The obvious answer was to carry a telescope up above the Earth's atmosphere, to get away from terrestrial 'interference', but the first serious attempt to do this was not made until 1954 when the intrepid

French astronomer, Audoin Dollfus, used a cluster of hydrogen balloons to climb to a height of 7,000 metres. In 1959, he reached 14,000 metres in a flight which in many ways called for as much courage and daring as that of the later American astronauts. During his two flights, Dollfus was able to measure the amounts of water vapour in the Earth's atmosphere at different altitudes and was later able to use this data to correct telescopic observations of Venus made from an observatory at 4,000 metres on the Jungfrau in Switzerland. He concluded that there was definitely a small amount of water in Venus' atmosphere, but he could still only observe the extreme outer part of it.

In 1964, Dollfus' work was refined by the American astronomer, J. Strong, who used an unmanned balloon to carry an automated telescope to a height of 26,000 metres. This gave similar results to Dollfus', and also suggested that the water in Venus' upper atmosphere was probably in the form of tiny ice crystals. The spectroscopic data also suggested that the clouds of Venus might be similar to the cirrus clouds – also composed of ice particles – that exist at great heights in the Earth's atmosphere.

Because the Earth's atmosphere contains only 0·03 per cent of carbon dioxide, it is much easier to use spectrometers to search for this gas in the atmosphere of other planets. As early as 1932, two American astronomers, Walter S. Adams and Theodore Dunham Jr, had shown that carbon dioxide existed on Venus in large quantities. Later, it became clear that it was much the most important component, though some astronomers thought that nitrogen might be abundant as well. Other observations suggested the presence of traces of hydrochloric and hydrofluoric acids and carbon monoxide, but there was little evidence of oxygen.

Thus, by telescopic observations alone, astronomers had learned quite a lot about the atmosphere of Venus, but the surface remained totally inaccessible. It was impossible to say anything about the surface temperature, or even the pressure at the base of the atmosphere. The fact that carbon dioxide was abundant, however, was important, and was used in some provocative speculations about the climate of Venus, of which more later.

The surface temperature: a sizzling surprise

In 1956, a team of radio astronomers began studying radio emissions from Venus. To their surprise, they found that the planet is a source of intense radiation at radio wavelengths, and, more important, that Venus radiates radio waves as though it were a hot body with a surface temperature of several hundreds of degrees centigrade. As with the early radio studies of Mercury, this observation conflicted with the prevailing preconceived ideas about what the surface temperature of Venus *ought* to be, so a number of rather elaborate hypotheses were invented to explain away the intense radio emission. These ranged from electrical discharges from liquid droplets in the clouds to charged particles trapped in a Van Allan belt like that around the Earth.

None of these ideas really held water – some of them were even self-contradictory – and they were rapidly disowned when the spacecraft missions forced scientists to accept that Venus really was an extremely hot place, and not the cool, tranquil haven that some of them wanted to believe in.

The rotation period: three centuries of debate

It was difficult to determine the rotation period of Mercury because that planet is always so difficult to observe. Notwithstanding this, Mercury does at least have visible surface features, though these did not help much in the determination of its rotation period. Venus is much easier to observe than Mercury, but its cloud-shrouded surface prevents measurement of its rotation period by simple means. It might seem obvious that it would be unwise to say too much about the rotation period of a body whose surface is as blank as that of a billiard ball – Cassini, in fact, said exactly this in 1666. Later astronomers were much less cautious, and some claimed that they had arrived at *the* rotation period.

In the eighteenth and nineteenth centuries, two alternative views were current: one suggested that Venus had a rotation period similar to the Earth's twenty-four hours; the other that Venus' rotation period was the same as its year, namely, 225 days. (This was authoritatively proposed by Schiaparelli, the astronomer who had also suggested that Mercury's rotation period was the same as its year. He unfortunately managed to be wrong in both cases.) Both these suggestions were sensible in that they were consistent with the state of knowledge at the time, but they were not much more than intelligent guesses.

At the beginning of the twentieth century, some serious attempts to measure the rotation period were made. The earliest of these involved examination of the Doppler shift of lines in the spectrum of light from Venus: exactly the same principle that was used with radio waves on Mercury (page 150). The first such study, carried out in 1956 by Robert S. Richardson of the Mount Wilson and Palomar Observatories, came up with the totally unexpected result that Venus rotates slowly in a retrograde sense; that is, it spins on its axis in the *opposite* sense to every other planet in the solar system. The errors in this determination, however, were large, and no one attached much importance to it.

Progressively more sophisticated observations of the same kind were made during the following years, and they all seemed to be pointing in the same direction: the rotation is retrograde, and very slow. In the 1960s, it was thought that the period was about four days. At about the same time, evidence was accumulating from a quite different source to confirm this supposition. Although Venus looks blank as a billiard ball when viewed in ordinary circumstances, it *does* have a distinctly blotchy appearance when photographed in the ultraviolet part of the spectrum. Once one can see features on a planet, of course, one has a means of measuring its rotation period. Fuzzy blotches had been seen on ultraviolet pictures of Venus as early as 1928, but no one had succeeded in determining a rotation period from them, partly because no one was expecting the rotation to be retrograde.

In the 1960s, when suggestions that Venus was a perverse planet were in the air, a French astronomer, C. Boyer, made a

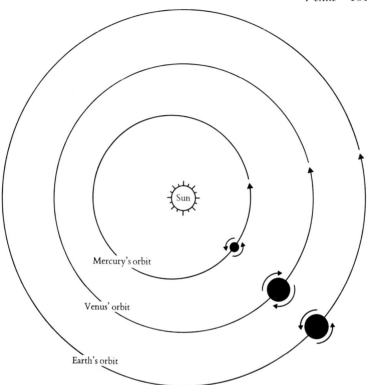

Figure 5.4. Venus' peculiar retrograde rotation. All the other planets rotate on their axis in the same anticlockwise sense as they revolve around the Sun. Venus' orbit is also anticlockwise, but its axial rotation is clockwise.

fresh study of Venus in the ultraviolet, and confirmed that the fuzzy markings did indeed show a consistent *four-day* retrograde rotation. It was appreciated at the time that these markings were caused by clouds in the atmosphere of Venus, and not by true surface markings, but the result was none the less a staggering one: Venus seemed to be unique in the solar system, and to contradict most of the ideas about the dynamics of planetary motion.

A further considerable surprise was in store. When radio astronomers got round to making detailed radar studies of Venus, they found, as they had hoped, that the radar beams penetrated Venus' thick atmosphere as though it simply did not exist, and that they could obtain radar echoes from the solid surface. This work was commenced in 1958. By 1962, it had been conclusively shown by tracking features across the surface of Venus that the rotation period was retrograde, and that *it takes 243 days to make one complete rotation.*

This is quite at variance with the ultraviolet data, but both observations are correct! The solid mass of Venus lumbers once round its axis every 243 days, but the atmospheric circulation is such that cloud formations whip round the planet in just over four days, at speeds of up to 100 metres per second. This may seem surprising at first, but it is not particularly outrageous – upper atmosphere winds on Earth often travel at high velocities relative to the ground surface. Terrestrial 'jet streams', however, are quite narrow, whereas the whole upper atmosphere of Venus seems to be moving at high speeds.

The 243-day rotation period conceals another elegant solar system statistic. The ratio between the Earth's period of revolution around the Sun (365 days) and Venus' rotation period is, as near as makes no difference, 3:2, and this means that every time the Earth, Venus and the Sun are lined up as in Figure 5.5, Venus presents the same face to the Earth. Again, as with Mercury, this is no mere coincidence. It is another example of spin-orbit coupling, and it underlines the beauty and harmony of the motions of the planets. It is not yet certain why Earth and Venus are coupled in this way, but some kind of tidal interaction may be involved.

Spacecraft missions to Venus

While it might seem appropriate for the Russians to have concentrated their energies on Mars, the 'Red Planet', and to have left Venus to the Americans, this is not quite how things have turned out. Although the Americans achieved a great success in 1962 with Mariner 2, which was the first ever 'fly-by' of another planet and

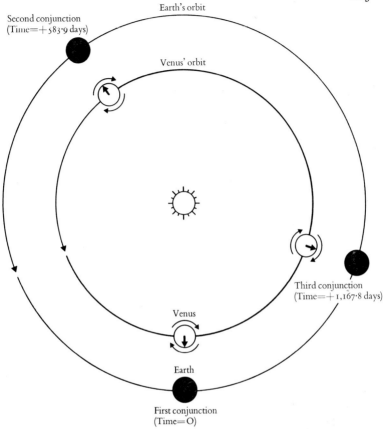

Figure 5.5. The Earth, Venus and the Sun lie in a straight line every 583·9 days. These occasions are known as conjunctions. At each conjunction, Venus presents the same face towards the Earth. At the third conjunction shown here, Venus will appear to have rotated on its axis six times, with reference to the Earth.

which obtained valuable temperature and magnetic data, Venus has been largely in the Russian sphere of influence. Their achievements in soft-landing spacecraft on the surface deserve the same recognition as the more widely publicized landings of the American spacecraft on Mars.

Figure 5.6. *The classic Mariner 10 view of Venus, which revealed for the first time the details of its atmospheric circulation. The picture was taken in ultraviolet light from a distance of 720,000 kilometres on 6 February 1974, while the spacecraft was* en route *for Mercury.*

Although ultimately successful, the Russian studies of Venus started in a depressing series of failures. The first launch was of Venera 1 in 1964. Communications failed almost immediately. The next attempt was Zond 1 in the same year. Communications failed after a month. Venera 2, in 1966, actually flew past Venus at a distance of 24,000 kilometres but failed to return any data. In 1966 again, Venera 3 scored success of a sort when it became the first spacecraft ever to reach the surface of another planet, but it also failed to send back any data.

No breakthrough was achieved until Venera 4, which, on 18 October 1967, penetrated the atmosphere of Venus and floated gently down to the surface on a parachute. During the ninety-four minutes of its descent, it transmitted a stream of data back about conditions in the atmosphere which were to be of first importance in planning later missions. It did not, however, apparently reach the surface in one piece. (Only a day later, the American space-craft, Mariner 5, successfully flew past Venus at a distance of only 4,000 kilometres, and provided a further mass of new data on the atmosphere.)

Another major advance came in 1969 with Venera 5, which survived the rigours of the descent through the atmosphere to become the first spacecraft to reach the surface in working order. It managed to keep working on the surface for only a very short time, however, before snuffing out. A similar fate befell Venera 6, later in 1969 – it went off the air almost immediately after reaching the surface. By now, of course, it was becoming clear that Venus was a pretty inconvenient place on which to put scientific instruments: the data were showing searing surface temperatures and crushing pressures. It was scarcely surprising that the space-craft could not survive on the surface. Imagine what would happen if you put something as simple as a transistor radio into the oven at home, with the temperature turned up to 'high', and tons of pressure thrown in as well!

Venera 7 was therefore more elaborately prepared than its predecessors. It managed to survive for fifteen minutes on the surface in 1970, and was able to radio back confirmation of surface temperatures of 477 °C, and pressures of 90 Earth atmospheres.

Venera 8 continued functioning for fifty minutes in 1972, and this time was equipped with an instrument like an exposure meter to measure light levels at the surface. Everyone supposed that Venus, apart from being horridly hot, would also be wrapped in Stygian darkness because of the thick blanket of clouds, and that surface photography would therefore be impossible without lighting. Venera 8, however, showed that sunlight does filter down to the surface of Venus, so that photographic studies would be worth while in future missions. Light levels are, in fact, comparable with those on Earth on a heavily overcast day.

The Venera 9 and 10 missions

The Venera 9 and 10 missions were conceived on roughly similar lines to the Viking missions which were launched in the same year. After the first eight Venera missions, the Russians had learned a great deal about the difficulties of spacecraft investigations of Venus, and this had enabled them to design a new generation of craft, better prepared to withstand the rigours of landing.

Like the Vikings, each Venera craft consisted of two components, one to orbit the planet and provide data on the planet as a whole, the other to soft-land on the surface. The main problem faced by the designers, of course, was that of *temperature* – the pressure of the Venus atmosphere is quite small compared to that at the floors of terrestrial oceans where studies are regularly conducted.

From their earlier missions, the Russians knew that they could expect surface temperatures at which materials begin to glow red hot. Delicate electronic equipment does not work terribly well when it is glowing red hot. The spacecraft were therefore thoroughly insulated, and refrigerating systems were built in to keep the sensitive components as cool as possible for as long as possible. To this end, the time that each craft spent in descending through the atmosphere of Venus was also kept as short as possible

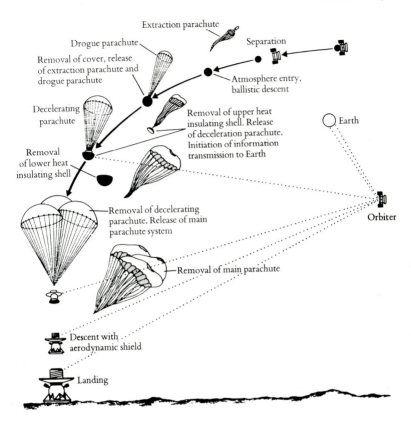

Figure 5.7. A schematic summary of the descent and landing sequence of the Russian Venera 9 and 10 spacecraft. (M. V. Keldysh)

by dispensing with the main parachute well *before* the craft hit the ground, and relying on the thick atmosphere to slow its descent sufficiently. A cleverly designed 'aeroshell' ensured that they finally struck the surface at about 7 metres per second, which is about the speed at which you would hit the ground if you jumped off a wall three metres high on Earth. An energy-absorbing framework was used to cushion the landing itself.

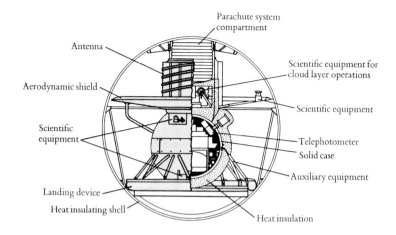

Figure 5.8.   General structure of the Venera landers. (M. V. Keldysh)

The view from the Veneras: the first pictures

Prior to the Venera 9 and 10 missions, the best-informed scientific opinions were that Venus would be a purgatorial planet where, during the Venusian noon, objects shaded from the direct light of the Sun would glow dull red from their own heat. This infernal prospect outclassed by far many science-fiction writers' most imaginative accounts of conditions on Venus. The first Venera pictures from the surface, although of great importance scientifically, did not live up to this vivid trailer. For one thing, the picture quality was not good, and furthermore, because of the techniques employed in scanning from the spacecraft, they had a distorted perspective which makes interpretation a little difficult.

The Venera 9 pictures, the first ever obtained from the surface of another planet, showed a bleak, grim terrain. A rock-strewn surface stretched unbroken to the horizon. Nothing else. By the time of Venera 9, no one expected to see anything animate on Venus, let alone shiny green Venusians riding around in flying saucers and talking with American accents, or even vanadium-

Figure 5.9. *Mankind's first views of the surface of a planet other than his own. The Venera 9 panorama is above; Venera 10 below. The pictures are best interpreted by looking first at the horizon on the right-hand side, and then rotating the whole picture to maintain the same sense of perspective.* (Picture provided by NSSDC)

eating swamp-living animals. But the sheer barrenness of the scenery dispelled finally and for all time any lingering illusions that Venus was a beautiful planet, a sort of 'alternative Earth'. People began to realize that Venus is one place in the solar system on which men may never land.

The view from Venera 10, which landed 10,000 kilometres away from Venera 9, was in some ways similar, in others surprisingly different. There was the same distorted perspective, the same sense of unreality, of desolation. The details, however, were quite different. Whereas Venera 9 was resting on a sloping surface of flattish, angular boulders, Venera 10 was resting on a flat, smooth rock surface with small particles of finer-grained material resting on it.

These contrasts were in themselves something of a surprise, since it had been supposed that Venus might be a rather uniform planet on which any topographic features would long ago have been smoothed away to monotony by erosion in the dense atmosphere. Such erosion, of course, would be enhanced if the atmosphere were corrosive, or if exceptionally strong winds prevailed at the surface, as some ideas suggested. The Veneras were equipped to investigate both of these, and other aspects of the Venusian environment. Venera 9 transmitted information for fifty-three minutes, Venera 10 for sixty-five. Both spacecraft were shut down by an automatic command before the temperature inside the spacecraft passed 60 °C, when malfunctions would begin to occur in the electronic equipment.

Surface experiments

The temperature and pressure on the surface of Venus had been well established by earlier missions. The Venera data confirmed the scorching temperatures, and indicated the same high pressures (90 Earth atmospheres) at both landing sites, showing that they must be located at about the same topographic height – about 1·5 to 2·0 kilometres above the mean surface. (Only rather crude topographic data were available from radar studies.) During their descent, the two craft made studies of the wind velocities in the atmosphere at different altitudes. The results (Figure 5.10) showed high velocities in the upper atmosphere, which was expected, but light winds – gentle breezes almost – at the surface. This has some important implications for erosional processes on Venus: if the winds are, indeed, always less than 1 metre per second, then wind erosion cannot play an important part in shaping the surface of Venus. However, less than two hours' worth of data from two stations 2,000 kilometres apart on the planet is not much to go on, so these results need to be treated with caution.

Venera 5 had already demonstrated that 97 per cent of the atmosphere of Venus consists of carbon dioxide. Veneras 9 and 10 were programmed to examine the all-important water-vapour content. They showed that, between 25 and 45 kilometres above the surface, only 0·1 per cent water is present, suggesting that Venus is a very dry planet indeed. This observation not only makes the possibilities of life existing on Venus vanishingly remote, but also means that we should not expect to find the kind of landforms that we are familiar with on Earth: the mountains, valleys and plains that we know are all largely the product of flowing water. With so little water on Venus, the chances of finding water-sculpted topography are practically non-existent; all the surface processes acting there must be 'dry' ones.

Veneras 8, 9 and 10 all carried gamma-ray equipment, capable of measuring the natural radioactivity of the surface rocks. By measuring the strength of the radiation at different wavelengths,

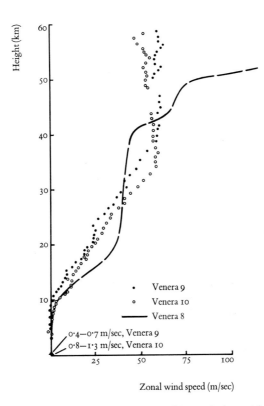

Figure 5.10. The wind speed on Venus, and its variation with altitude, as determined by the Venera spacecraft. (M. V. Keldysh)

it was possible to say something about the abundance of the principal radioactive elements, potassium, uranium and thorium. The rocks on which Venera 8 landed – of which there are no pictures – gave results that are comparable with the abundances of potassium, uranium and thorium in terrestrial *granitic* rocks. This is an important observation, since granitic rocks on Earth are characteristic of the continents: it suggests that there might be 'continental' material on Venus.

The gamma-ray data from the Venera 9 and 10 sites were quite different, and much more in keeping with what was expected. The data showed that the rocks at the two sites were broadly similar to terrestrial *basalts*, the rocks which make up the ocean basins. Basalts appear, in fact, to be some of the most abundant rocks in the solar system. The gamma-ray equipment was also capable of measuring the density of the surface rocks. The data again corresponded with basalt.

A closer look at the surface

Although the panoramas taken by the Venera spacecraft are rather fuzzy, they contain a good deal of information, some of which is not immediately obvious, and they raise some difficult questions. What is the material the spacecraft are standing on? How did it get there? How old is it?

The Venera 9 picture has well-defined boulders extending as far as the right-hand horizon: this shows that the horizon is relatively near, and suggests that the 'horizon' is not a true horizon but the crest of a slope. This is the first important observation: there are quite steep slopes on Venus. This may not sound particularly earth-shattering as a discovery, but it is important when one comes to think about *how* such slopes might form.

An inclinometer on the spacecraft in fact showed that it was heeled over at an angle of 30 degrees. In part, this was because one foot of the craft was resting on a boulder, but it is thought that the general slope in the area is about 20 degrees. This represents a gradient of about 1 in 3, too steep for a car to drive down comfortably. The picture clearly shows that the rocks resting on this slope are large, angular, flattish slabs, with rounded edges. They look, in fact, a bit like the fragments one would find in terrestrial scree slopes, and therefore, like scree material, they must be relatively youthful. Whatever erosional processes have been at work to break off the fragments in the first place, the rocks have not been carried far from their place of origin.

Venera 10 came to rest on a flat surface; it seems to be resting directly on top of a smooth, flat-topped rock outcrop. Closer study shows some small step-like features on the outcrop, reminiscent of those formed when layered terrestrial rocks are fractured. The main outcrop is surrounded by smaller ones, separated by darker, fine-grained material; angular slabs seem to have broken off from the outcrop and drifted sideways away from them. The main outcrop certainly has obvious fracture lines in it, but it is not easy to see how the sideways movement could have taken place, especially since it is impossible to resolve the details of the fine-grained material.

Both the Venera 9 and 10 panoramas show clear evidence of erosion and transportation. In Venera 9, angular boulders have been derived from some unknown source and are migrating down a steep slope. Here we are looking at a process that the Russian planetologists have called *decimetre scale degradation*. In Venera 10, the edges of outcrops are being rounded, slabs have apparently broken off from the main outcrop, and fine-grained material appears to be accumulating between the solid rocks. This the Russians have called *centimetre scale degradation*.

But how does this degradation happen? We have seen that there can be no water around to carry out the kind of erosion that we know on Earth. So could it be caused by the Venusian winds?

In theory, Venus is an excellent place to expect wind erosion. The atmosphere is so dense that the minimum wind speed required to lift a particle of given size is about one tenth that on Earth. The ultraviolet pictures of the planet show evidence of violent winds in the upper atmosphere. So, if there were violent winds at the surface, the effects would be spectacular.

But all the evidence from the Venera spacecraft shows that the winds on the surface of Venus are gentle. The prevailing winds of less than 1 metre per second would have little ability to transport material in suspension, to sand-blast rocks and erode them, or to build up dunes. It is, of course, possible that violent surface winds do blow from time to time on Venus, but, on the whole, Venus seems to have rather boring weather. Because its rotation axis is nearly perpendicular to its orbital plane, there are no seasons to speak of. (The Earth's axis is inclined at $23\frac{1}{2}$ degrees, remember.) There are practically no daily variations in temperature because the rotation period is so slow and the atmospheric blanketing so complete. There do not even appear to be marked differences in conditions with latitude. Thus, the poles of Venus appear to be nearly as oppressively hot and nasty as the rest of the planet.

Although Venus' atmosphere does not seem conducive to balmy breezes, it does have some intriguing properties. If an atmospheric disturbance could be initiated, and particles swept up into it, the density is such that the particles would stay in suspension for a long time before settling out, and this opens out one mechanism of erosion that does not take place on Earth.

Consider what happens when you stir up the mud on the sloping sides of a muddy pool with a stick. As a rule, the thick opaque soup of suspended mud blocks out your view completely. (A dust storm on Venus?) But if you look carefully, you may see tightly convoluted little clouds of sediment rolling off in long trains directly downslope, hugging the bottom, and with opaque curtains of sediment-laden clouds above them. These are known as *turbidity currents*. They are well known in the oceans of the Earth, and their powerful erosive powers have been well documented on occasions when such currents have swept down the continental slopes of the Atlantic into the deep ocean at speeds of 50 kilometres per hour, and carried away underwater telephone cables.

Now the density of the atmosphere on Venus is so great and the settling velocity so small, that dust clouds in it might behave in the same way as sediment particles in water on Earth. So long as there are differences in topographic elevation, and sufficient quantities of fine-grained material to hand, any atmospheric disturbance on Venus could trigger off a 'turbidity current' which would sweep downslope like a hurricane, and would be imbued with violently erosive powers.

The nature of the rocks

Although the chemical data obtained by the Venera spacecraft provide some information about the composition of the rocks on Venus, they don't tell us much about how they were formed. There are several possibilities.

The most obvious one is that they are lavas, erupted from a Venusian volcano. This is strongly suggested by the basaltic gamma-ray data, and it conforms with the observation that basalt lavas are common in the solar system. Against this, however, is the fact that both the Venera 9 and 10 pictures suggested that the rocks are *layered*, and layering simply does not occur in lavas – it is much more typical of sedimentary rocks. Unfortunately, the quality of the Venera pictures is not good enough to show exactly what gives the rocks their layered appearance. If distinct bands of different minerals are present, then one can effectively rule out lavas, but if the apparent layering is caused by the presence of horizontal *joints* or fractures in the rocks, then lavas are still in with a chance. Many examples of terrestrial lavas are known which have a 'platey jointing', often parallel with the ground surface, which may be a result of rapid chilling of glassy lava. In a few cases, this jointing is sufficiently intense and closely spaced to cause the rock to break up into thin slabs like roofing slates.

Although basalt lavas are the first rocks that come to mind, there are many other kinds of rock that have basalt compositions. There are some intrusive rocks (called gabbros) which crystallize deep below ground rather than on the surface, and these occasionally have well-defined layers within them. And, of course, volcanoes do not always erupt lavas – some eject huge quantities of ash high into the atmosphere. On Venus, ash erupted from a volcano could be held in suspension for a long while before coming to rest on the surface, and it is possible that the whole of the planet is covered with layers of volcanic ash. If this became hardened or lithified, it might look like the material shown on the Venera pictures.

Another possibility is that the rocks are the remains of impact melts; solidified remnants of material ejected and melted by violent meteorite impacts on Venus at some point in the past. All the planetary surfaces that we can observe show evidence of meteorite bombardment, so Venus must have suffered in the same way. It is also just possible that the rocks are true layered sediments deposited in Venus' geological past, lithified and brought to the surface by erosion. But Venus is an exceedingly dry place, and it is difficult to see how such sediments could have formed in the absence of water.

It seems, then, that while there is a strong likelihood that the rocks we see on Venus are basalt lavas, there is not sufficient evidence to confirm this unequivocally, and the issue will probably remain in doubt until further landers have carried out more complete analyses. There is one other line of evidence, however, that provides us with data on the surface geology of Venus.

Radar studies made both with powerful Earth-based radio telescopes and orbiting spacecraft have been able to map out some of the surface features. The overall impression created by these studies is rather boring, unfortunately – huge parts of the planet seem to consist of monotonous, gently rolling plains. There is little relief, and few features are more than 3 kilometres high. There are some tantalizing hints of some much more exciting features, however. A large, circular feature known only as *beta* has been interpreted as an enormous shield volcano, perhaps 800 kilometres in diameter. Because the resolution of the imaging system is so poor (only 25 kilometres), it would be unwise to assume the existence of this volcano to be firmly established, but, *if* it exists, it will be far and away the biggest in the solar system. A giant rift valley 7 kilometres high and several hundred kilometres long has also been inferred from radar data. If this structure exists, it will be a pointer towards the operation of plate tectonics on Venus, a thought to make terrestrial geologists' mouths water.

While the volcano and rift valley must remain in doubt, there are certainly many large crateriform structures on the surface, comparable with the large craters and ring basins on the Moon and Mercury. These, however, appear to be extremely low, flat features. One of the best-known craters is 160 kilometres in

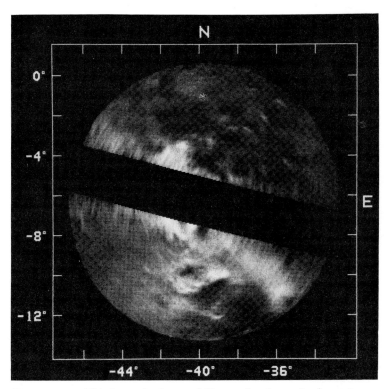

Figure 5.11. *This is a radar image of Venus, showing a circular region some 1,500 kilometres in diameter near Venus' equator. The southern part of the image contains a large 'crater' about 75 kilometres across and 500 metres deep, which may be of either volcanic or impact origin. The black strip is an area for which data is not available.* (NASA Jet Propulsion Laboratory)

diameter, yet its rim is only 500 metres high. This very subdued topography could be the result of extremely intensive erosion taking place beneath the thick Venusian atmosphere. The atmosphere has also effectively blocked out much of the meteorite bombardment of the planet: small meteorites simply cannot penetrate it without burning up. Hence no small impact craters are found on the surface of the planet.

In possessing large craters, Venus resembles the Moon and Mercury. For all its similarity with the Earth, however, the radar data do *not* show the existence of separate 'continental' and 'oceanic' areas. These two observations together suggest that Venus may *not* be a dynamic, active planet like the Earth, with plate tectonic processes constantly changing its surface. If this is correct, then the basaltic rocks on the surface may have originated in the early history of the planet as the result of cataclysmic impact events like those that gave birth to the lunar maria.

Pioneer Venus: a shower of spacecraft

Anyone unwise enough to be living on Venus in December 1978 might have been justified in thinking that he was seeing the beginning of some new War of the Worlds: seven separate spacecraft reached the surface within a few days of each other. Two of these were Russian, constituting the Venera 11 and 12 missions, and five were American, forming part of a complicated mission known as Pioneer Venus.

Veneras 11 and 12 both landed successfully, transmitting data from the surface for 95 and 110 minutes respectively, but no new pictures were obtained, and it seems that at least some of the instruments on board did not work satisfactorily. Perhaps the least expected discovery that the Veneras turned up was that of thunderclaps on Venus – one enormous discharge is reported to have rolled around Venera 12 for no less than fifteen minutes! Clearly, Venus is no place for ladies who dive for cover beneath the dining room table at the first distant rumble of thunder.

The Pioneer Venus spacecraft were designed specifically to investigate Venus' atmosphere rather than the surface. Pioneer Venus 1 was a straightforward spacecraft, which was placed into an elliptical orbit around the planet, and the orbit was gradually trimmed so that the craft brushed nearer and nearer the atmosphere. Eventually, it reached a minimum height of 150 kilometres, which enabled it to dip briefly into the upper atmosphere on each orbit, sampling it directly.

Pioneer Venus 2 consisted of a 'bus', one large probe and three small ones. The 'bus' was spun up to about 48 r.p.m. before encountering Venus, so that the four probes were flung off by centrifugal force on a command from Mission Control. This ensured that each probe would follow a different trajectory towards Venus, and arrive at a different part of the planet. The large probe carried a parachute to slow its descent through the atmosphere and to ensure that it would reach the surface in working condition, though it was *not* intended to survive the landing. The three smaller probes had only their blunt, aerodynamic profiles to slow them down. They hit the upper fringes of Venus' atmosphere, travelling at 11 kilometres per second, but soon reached dense enough atmosphere to experience truly massive deceleration – no less than 565 g's momentarily. Such forces would turn any human occupant of the spacecraft to jelly in an instant, but the probes not only survived this, and kept on working throughout their journey through the atmosphere, but one of them, amazingly enough, even survived impact with the surface, and kept on transmitting data for sixty-seven minutes before it was snuffed into silence by the surface temperature.

Each small probe weighed only 90 kilograms. Of this, only 3·5 kilograms consisted of scientific instruments, the rest being taken up with the aerodynamic heat shield, insulation, and the spherical titanium pressure vessel required to protect them. Also on board was an 11-amp-hour 28-volt silver zinc battery to power the instruments and the radio transmitter. An 11-amp-hour battery is not very powerful; it would be too small for even a small motor-car. None the less, radio signals from the probes were successfully picked up on Earth, where the signal strength was only 10^{-20} watt. This is an almost unimaginably weak signal, millions of times weaker than the light of a glow-worm seen across the length of a garden. Detecting such feeble signals required extremely sensitive equipment, and elaborate precautions to avoid interference. At one of NASA's Deep Space Network receiving stations in Chile, the army sealed off the main arterial road in the country to prevent stray signals from automobile ignition systems cutting across the faint streams of data coming in from Venus.

The atmosphere: facts stranger than fiction

Many separate experiments were carried out on Venus' atmosphere by the Venera and Pioneer spacecraft. Their results have been more than adequate to confirm the graphic prediction that Venus 'seems very much like the classic view of hell'. Indeed, hell would be a pleasant holiday resort compared with Venus and its corrosive clouds and lethal fumes.

The data confirmed that the bulk composition of the atmosphere is similar to what had been indicated by the earlier Veneras: 95 per cent carbon dioxide, with a few per cent of nitrogen and small traces (especially near the surface) of oxygen, sulphur dioxide (tear gas) and water vapour, the latter forming only a few tenths of 1 per cent. Such an atmosphere would be suffocating to humans, if not outright poisonous; but of much greater interest to planetary scientists were data on gases present in only trace quantities. They discovered that Venus' atmosphere contains far more of two isotopes of argon – argon 36 and argon 38 – than does the Earth's, but about the same amount of argon 40.

Now, argon 40 is produced by the radioactive decay of a potassium isotope, one of the isotopes present within the Earth, and is partly responsible for its hot, mobile interior, and thus for the processes of volcanism and plate tectonics. By implication, then, Venus should have a similar amount of radioactive potassium in its interior, which is also likely to be hot and mobile.

The argon 36 and 38 are much more difficult to account for. They are several hundred times more abundant on Venus than on the Earth, and since they *cannot* be manufactured by radioactive decay, their presence can only relate to conditions prevailing during the initial formation of the planet. Some scientists have argued that Venus may have been able to retain more of its primitive atmosphere, derived directly from the primitive solar nebula, than the Earth. In the next few years, when there has been more time to mull over the data, other ideas will probably be forthcoming. It seems certain, however, that the anomalous argon results will be the most important discovery of the 1978 season of exploration of Venus.

Important though the argon data are, the clouds on Venus are much more obvious features of its atmosphere. The probe data showed that there are three well-defined 'decks' of clouds, separated by clear layers. The top deck is located at an altitude of about 65 kilometres and is probably itself several kilometres thick. It is this deck of clouds which is responsible for the windswept appearance of Venus in the ultraviolet. The violent winds which carry the clouds round Venus in only four days seem to be confined to this altitude. At lower altitudes, wind speeds are much lower, and at the surface only the gentlest of zephyrs has been recorded. 'Cloud' is probably too strong a word for the upper deck, which seems in fact to be more like a thin terrestrial haze. It appears to be made of tiny liquid droplets about 1 micron in diameter (1 micron is one thousandth of a millimetre).

At about 58 kilometres altitude is another haze layer, less dense even than the uppermost one. Here, some very large particle sizes were encountered, up to 30 microns. These have been interpreted as being tiny *sulphur* crystals suspended in the atmosphere. Visibility in this layer might be comparable with what one sees on Earth during a summer heat wave, when anticyclonic conditions lead to the build-up of a pall of smoke from chimneys and stubble fires, automobile exhausts, dust and general pollution. All of these are unpleasant enough, but at least we do not have particles of solid sulphur in our atmosphere! The presence of the sulphur helps to explain another feature of telescopic observations of Venus: the clouds appear not to be perfectly white, but tinged with lemon yellow.

The bottom layer of clouds is the thickest. It seems to have a well-defined base at about 49 kilometres above the surface, and to be about as dense as ordinary clouds on Earth, with visibility within it restricted to less than 1 kilometre. It appears to consist of liquid droplets of a wide range of sizes. When it passed through this cloud layer, one of the Pioneer probes reported the presence of a shower cloud immediately beneath it. The most likely constituent of the drops in the shower seems to be sulphuric acid . . .

Thus, whereas we on Earth can lie on our backs on a fine summer day and enjoy the relaxing sight of fleecy white clouds of pure water vapour chasing each other across the sky, anyone unfortunate enough to live on Venus would have to put up with a constant leaden-grey overcast sky, from which showers of 80 per cent sulphuric acid occasionally descend. A depressing prospect.

Hot, concentrated sulphuric acid is violently unpleasant stuff, as many readers may recall from their school days. An even more unpleasant possibility exists, however. There is some evidence for the existence of both hydrogen chloride and hydrogen fluoride in the atmosphere of Venus. The latter could react with sulphuric acid to form fluorosulphuric acid (HSO_3F). This has been described as the most corrosive fluid in the solar system. It is capable of dissolving sulphur, mercury, tin, lead and most rocks. If a rain of this acid were to reach the surface, its erosive powers are best left to the imagination . . . Next time you are struggling with a recalcitrant umbrella on some windy street corner, be grateful that the rain slashing down is not hissing on the pavement, corroding the cobblestones and dissolving the drain pipes.

Greenhouses and runaway greenhouses

Much of this chapter has dwelt on the unpleasantnesses of Venus' atmosphere, without going into the reasons. Many major questions need to be answered: why is the atmosphere so massive? Why are temperatures so searingly high? Why is there so much carbon dioxide, and so little water? Why is Venus so totally different from the Earth? Not surprisingly, all these questions are inseparably intertwined.

The massive carbon dioxide-rich atmosphere is not as anomalous as it may seem at first sight: there is only as much carbon dioxide on Venus as there would be in the Earth's atmosphere if all the carbon dioxide contained in terrestrial rocks, such as limestones, were liberated. This raises a nightmarish possibility: Earth could end up as the same stifling hell as Venus.

Because Venus is nearer the Sun than is the Earth, one might reasonably expect to find higher temperatures there. Things are not as simple as this, however. For one thing, the temperatures are *much* higher than could be explained simply by the planet's closeness to the Sun. For another, the sheer brilliance with which Venus shines in the night sky indicates that the clouds in its atmosphere must reflect a very high proportion of the Sun's rays – about 80 per cent, in fact. Why, therefore, is the surface so hot?

This is where the greenhouse comes in. When the Sun's rays shine onto a greenhouse, most of them pass through the glass unobstructed, but the infra-red wavelengths are partially blocked. Now, when the sunlight falls on the floors, walls, shelves and flowerpots in the greenhouse, some of it is reflected (this is what you actually see); but some of it is absorbed and reradiated at *different wavelengths*. This is the whole point of a greenhouse: much of the reradiated radiation is in the infra-red wavelengths; it is heat. Because the glass of the greenhouse is partially opaque to the infra-red, this heat is trapped inside. Hence the temperature rises. Tender blossoms thrive. Gardeners are happy. Of course, this 'one-way' filter is not very efficient. If it were, the temperature in the greenhouse would rocket, the tender blossoms would be carbonized in their pots, and the gardeners would be anything but happy.

Venus appears to be an example of an efficient greenhouse. We've seen that some radiation from the Sun does reach the surface. Much of the incoming radiation is blocked by the clouds, and almost all of the reradiated infra-red wavelengths are blocked by them. The 'glass' in the Venusian greenhouse is simply the atmosphere. Carbon dioxide is extremely good at absorbing infra-red radiation, and so is sulphuric acid. Much the best absorber, though, is water vapour. Although there is only a tiny amount of water in Venus' atmosphere, it is sufficient, combined with the other gases, to ensure that surface temperatures on Venus are maintained at their purgatorial levels.

It is not difficult to appreciate just how good a greenhouse water vapour is. Clear nights on Earth are always the coldest. Even in hot deserts, night-time temperatures often fall below zero when the sky remains free of cloud. The ground, heated by the Sun during the day, simply radiates all its heat out into space, with nothing to stop it. If the sky is overcast, by contrast, the night-time temperature may be ten or twenty degrees warmer: the water vapour in the clouds blocks the reradiated heat, and stops it going to waste in space.

The high temperatures on Venus, then, can be satisfactorily explained by the greenhouse effect. But how did it get its atmosphere in the first place? This raises a whole set of separate problems, not all of them soluble at present, as the peculiar argon abundances clearly demonstrate. It seems likely that Venus may have started off 4·6 billion years ago with a fairly thin, cool atmosphere. Then, as gases were blown up from the interior by volcanic activity, more and more water began to accumulate at the surface. Just possibly, this could have formed lakes and oceans. Then, as more and more water vapour accumulated in the atmosphere, the greenhouse effect got under way and surface temperatures rose. As temperatures rose, so more water was evaporated off, until a 'runaway greenhouse' was under way.

Things did not reach equilibrium until *all* the water on Venus was in the atmosphere, and *all* the carbon dioxide in carbonate rocks and minerals had been driven off. Venus may have resembled an enormous lime kiln at that time. This is where the uncomfortable parallel with Earth comes in. If the Earth were a little nearer the Sun, or if the Sun were to get a little warmer – not an impossibility over geological periods of time – then temperatures on Earth would rise, more water would enter the atmosphere, the temperature would rise more, the oceans would boil dry, and a 'runaway greenhouse' would ensue. As on Venus, this would only reach an equilibrium when all the carbonates in limestones and the like had been turned to lime, and carbon dioxide added to the atmosphere.

There is, fortunately, one factor that could possibly help to prevent this unpleasant ending for the world, and that is that Earth has *much* more water than Venus. While we should all be fervently grateful for this, it is difficult to see *why*. Earth and Venus are basically similar planets; they formed near one another in the primitive solar nebula; therefore they should have broadly similar amounts of water. Because it formed closer to the Sun, it is possible that Venus simply was not endowed initially with as much water. It is also possible that it originally had more, but has somehow lost nearly all of it. This problem is bound to remain one of the most interesting and important in planetary science for many years to come.

The interior of Venus

Because its size, mass and density are so close to that of the Earth, it is reasonable to suppose that Venus has a similar internal structure to that of the Earth: namely, a metallic core, a dense silicate mantle and a lighter crust.

One of the most outstanding features of the Earth, considered as a planet, is that it has a hot, mobile mantle, and a crust thin enough to break up into large plates which shift around relative to one another. The fact that the radar data show that large impact structures are present on Venus, means that Venus is unlikely to have such vigorous crustal and mantle processes. If continental drift and plate tectonics are as active on Venus as they are on Earth, then the large impact structures would have been obliterated long ago. Some of the radar data, however, suggest that *some* internal processes are at work. To resolve this issue, what is needed are detailed radar images of the planet from an orbiting spacecraft. Such a mission is already at an advanced state of planning; it is called the Venus Orbiter Imaging Radar mission, and will be flown in the 1980s. The images, when they are obtained, will be as exciting as the first close-up pictures of the Moon; Venus might be even more exciting geologically than the Earth.

One of the consequences of the Earth's having a large, molten iron core is that motions of the molten metal generate a powerful magnetic field. One of the first things that spacecraft missions to Venus did was to look for a comparable magnetic field around Venus. They found that Venus has *no* intrinsic magnetic field. This observation does not necessarily mean that Venus does *not* have a molten metal core. Unfortunately, not much is known about the origin of the Earth's magnetic field, as we saw in

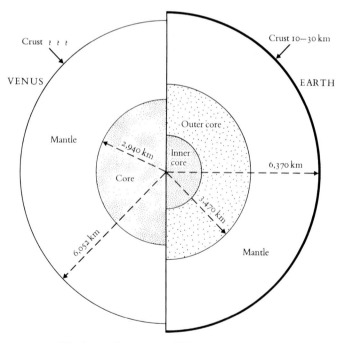

Figure 5.12.   The internal structure of Venus, compared with the Earth.

discussing Mercury's field (pages 167–8), but it does seem as though the Earth's rotation has something to do with it. The core may rotate at a slightly different speed from the mantle, and this may generate motions in the core that act as a dynamo which generates the field. Since Venus rotates extremely slowly on its axis, it is possible that this kind of differential movement does not take place, and therefore that the 'dynamo' never gets switched on.

Assuming that Venus does have a large, fluid metallic core, can we be sure that it is like the Earth's? The answer to this seems to be no. Although the similarities between Earth and Venus in size and density have been stressed in this chapter, Venus is actually about 2 per cent less dense than the Earth: a small but significant difference. This could have resulted from small differences in the process of formation of the two planets from the original solar nebula, which led to Venus getting a smaller share of metal than the Earth. There are, however, grounds for thinking that this primary differentiation is unlikely, and that secondary, *chemical* processes are responsible. If more of the iron on Venus were present as iron oxides in the mantle than in the Earth, there would be less iron in the core and the density of the planet as a whole would be slightly lower. ·

The large-scale oxidation of iron on Venus might result from its having formed at a higher temperature than the Earth. This is not too surprising, in view of its nearness to the Sun. However, it is also possible to argue that exactly the *opposite* situation might prevail: that there is *no* oxidized iron in Venus. In this case, the difference in density between Venus and Earth could be satisfactorily explained if Earth had more sulphur in the core than Venus.

Venus, then, has proved not to be the idyllic planet that the early writers may have hoped. Yet, although much of its romantic appeal may have been blown away in a blast of spacecraft exhaust gases, the planet has emerged as one of the most fascinating in the solar system. Since it is also the easiest for spacecraft to reach, we may be sure that a great many new discoveries will be made over the next two decades or so.

6. Mars – the Abode of Life?

From the dawn of civilization, the fiery, blood-red colour of Mars has been associated in mens' minds with virility and militarism. It is no coincidence that the symbol used by the ancient astrologers to denote Mars was the 'shield and spear' ♂, and that the same sign is now used by biologists the world over for the male sex. The links between Mars and the carnage and destruction of war extend back as far as the Sumerian civilization which flourished in the Middle East *c.* 3000 B.C. Their successors, the Chaldeans, formally adopted Mars as their god of war and christened it *Nergal*. The Greeks and Romans followed suit, but called the planet *Ares* and *Mars* respectively. It is the Roman name which has lived on into European languages, and we still pay annual tribute to their god of war – in the month of *March*.

The early religious and astrological studies of Mars propagated a fertile crop of myths and legends about the planet; sadly, it would be out of place to explore these here. The scientific study of Mars, however, has proved wonderfully rewarding, and has given birth to a complete second generation of myths, some of which are discussed in this chapter.

The first serious contribution that Mars made to the world of science came even before the first telescopic observations of it were made. It is obvious to even the most casual watcher of the skies that the appearance of Mars in the night sky changes considerably over the course of years. In one year – as in 1971, for example – the planet glares down fiercely, an unmistakable object, brighter than any other planet except Venus. A year or two later, and the planet is a most unimpressive sight, scarcely brighter than some of the red stars.

Furthermore, as any reader of astrologers' columns knows, Mars is constantly changing in position relative to the familiar constellations of stars. The apparent motion is not steady, but erratic. Sometimes Mars seems to be moving one way; later it appears to turn round and move off briefly in the opposite direction.

It is easy now for us to appreciate the reasons for these variations: they arise because of the changing relative positions of Mars and the Earth as they orbit around the Sun at different speeds and distances. It will be recalled from Chapter 1 that it was Johannes Kepler who first worked out the ways in which planets move. For him, it was far from easy to explain the variations in the motions of Mars and the other planets. Not only did he have to reconcile Tycho Brahe's mass of accumulated observational data with his theories, but he also had to overcome his own deep convictions, and those of all his contemporaries, about the ways in which planets *ought* to move.

The elliptical shape of Mars' orbit around the Sun means that its distance from the Earth varies considerably. At its closest, it may be only 55 million kilometres away; at its furthest, it may be nearly eight times that distance. Even at 55 million kilometres, Mars is still extremely distant, and it is this fact which gave birth to many of the myths and misconceptions about it. It has a diameter of only 6,787 kilometres, which is small by planetary standards. This means that, even with a good telescope at the times of its closest approach, observing Mars is a bit like trying to make out details on the Moon with the naked eye. And, of course, the high magnification necessary exaggerates all the restless turbulence in the Earth's atmosphere.

It is not surprising then, that the earliest observers, using crude telescopes, were led to some rather dubious conclusions. Their work, however, was remarkably good and, more important, *objective* in comparison with that of some later astronomers, who, armed with the best telescopes that money could buy, added to

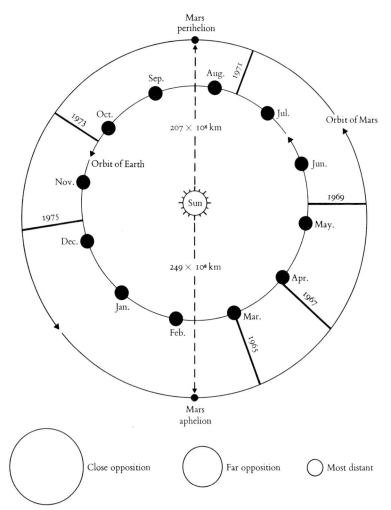

Figure 6.1.   The geometry of the orbits of the Earth and Mars is such that on some occasions, such as August 1971, the distance between the two planets reaches a minimum of 55 million kilometres. In other years, such as 1965, the two planets never come closer than 100 million kilometres. The circles show the relative apparent sizes of Mars as seen from the Earth on these occasions, and at its most distant.

their technological equipment a few intellectual filters to help them observe the features they wanted to see.

It is not known who first turned a telescope on Mars, or what he saw. The first observations of which there is a record were made by an Italian astronomer, Francisco Fontana, in 1636. The drawings he made are not very informative; they seem to be a record only of the defects in his telescope, which caused him to see a dark spot at the centre of the planet, and a dark ring round the outer part.

Telescope design improved steadily throughout the seventeenth century. By 1659, Christian Huygens, a Dutch physicist, had improved the performance of his own telescopes so greatly that

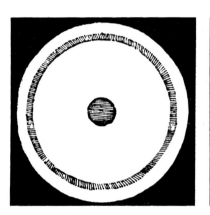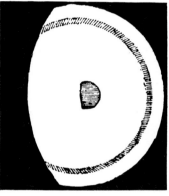

Figure 6.2.   Two of Fontana's early telescopic views of Mars. Note the similarities between these and his drawings of Venus (Figure 5.2, page 175).

Figure 6.3.   Huygens' sketch of Mars, showing the unmistakable, triangular Syrtis Major.

he was able to make out dusky surface features on Mars, and his sketches show one that can still be identified today: the roughly triangular-shaped feature known as the Syrtis Major. With such distinctive markings to use as reference points, Huygens was then in a position to determine the period of rotation of the planet. He found this to be just over twenty-four hours, nearly the same as the Earth's.

In 1666, Giovanni Cassini made some more precise observations and found a value for the Mars day of about 24 hours 40 minutes, which is reasonably close to the presently accepted value. Cassini is also credited with the discovery of the famous polar ice-caps on Mars, which were to figure so prominently in the later science-fiction stories about the planet.

It was not until about a hundred years later that speculations on the eternal question of 'Is there life on Mars?' really got under way. Towards the end of the eighteenth century, the German-born astronomer, William Herschel, made the important observation that the rotation axis of Mars is inclined at an angle of nearly 24 degrees to its orbital plane, similar to the Earth's $23\frac{1}{2}$ degrees. Unfortunately, he used this fact to argue that Mars must have four seasons just like the Earth's, and therefore, since the Martian 'day' is nearly the same length as the Earth's, that 'the inhabitants of Mars probably enjoy a situation similar to our own'.

Since it was then commonly accepted that most of the planets were inhabited, Herschel's authoritative views reinforced this general belief and, in particular, gave support to the idea that there was intelligent life on Mars.

Mapping Mars

During the nineteenth century, Mars came under increasingly close scrutiny as the quality and size of telescopes increased. By the 1830s, two Germans, Wilhelm Beer and Johann von Madler, had produced the first map of Mars. Although it was poor by present standards, they had attempted the job seriously and had

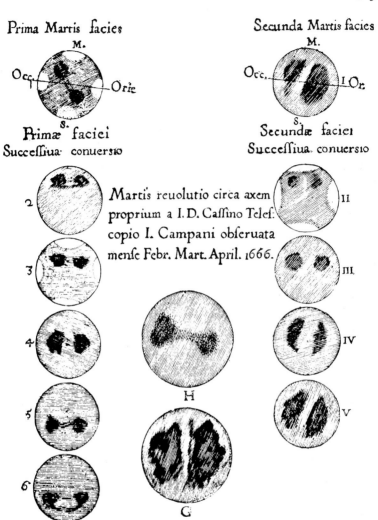

Prima Martis facies

M.

Occ. — Orie

S.

Primæ faciei
Succeſſiua conuersio

Secunda Martis facies

M.

Occ. — I Or.

S.

Secundæ faciei
Succeſſiua conuersio

Martis reuolutio circa axem
proprium a I. D. Caſſino Teleſ
copio I. Campani obſeruata
menſe Febr. Mart. April. 1666.

2

3

4

5

6

H

G

II

III

IV

V

*Figure 6.4.   A series of drawings of Mars made by Giovanni Cassini in
February and March 1666. The polar caps are marked, and, in the
left- and right-hand columns of sketches, the appearance of each
hemisphere after successive rotations of the planet is shown.*

established a set of lines of latitude and longitude as a reference grid. The basis of this grid, and in particular its prime meridian, is still in use today.

After this pioneering work, many other nineteenth-century astronomers spent long and frustrating hours at their telescopes, attempting to make more elaborate maps of Mars. Many of their problems stemmed from the fact that even the best telescopes could not show fine details. The only really conspicuous features were the white polar ice-caps; the rest of the surface markings were dusky and indistinct. These features were difficult enough to see at the best of times, because of the turbulence in the Earth's atmosphere, but, to make matters worse, the markings never appeared to be exactly the same twice. Some astronomers thought that these variations were caused by great dust storms sweeping across the desert surface of the planet – not a bad guess, incidentally – while many others thought that they showed seasonal variations in the vegetation cover which they thought was growing over large parts of the surface. The changes in appearance caused havoc for the mappers, since no two maps of Mars ever looked alike and, in many cases, names were given to places which simply did not exist.

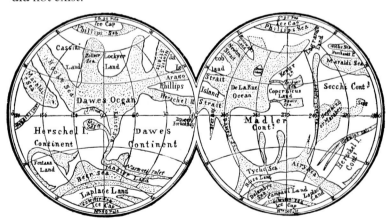

Figure 6.5.   Richard Proctor's remarkable map of Mars.

Richard Proctor, an English astronomer, made the first stab at finding names for Mars in 1867. He followed the same general principles that had been used on the Moon, and named things after famous observers of the planet. Kepler, Tycho Brahe, Madler, Cassini and Herschel were all duly honoured, and no one could take exception to these distinguished scientists. But Proctor seemed to have a bee in his bonnet about a certain English astronomer, the Rev. William R. Dawes. Why Proctor felt so strongly about this relatively obscure cleric we shall never know, but he went to the length of naming no less than six major features on Mars after him: Dawes Ocean, Dawes Continent, Dawes Sea, Dawes Strait, Dawes Island and Dawes Bay. Not surprisingly, Dawes was not as universally popular among other astronomers as Proctor might have supposed, and Proctor's names were soon quietly forgotten.

It was left to an Italian with a taste for Latin to advance the first generally accepted nomenclature for maps of Mars. In the summer of 1877, Giovanni Virginio Schiaparelli drew a brand-new map, and carefully plotted sixty-two major features, which he named with apparent relish. The light areas, called lands or continents by poor old Proctor, he named after real or mythical terrestrial countries – Arabia, Syria, Arcadia, Utopia and so on – while the dark areas he mapped as 'seas', and gave them some splendidly polysyllabic names which roll satisfyingly off the tongue: Mare Tyrrhenum, Aurorae Sinus, Aonis Sinus, Margaritifer Sinus and so forth.

Schiaparelli is most famous, however, as the discoverer of the so-called Martian 'canals', though these had first been described by an Italian priest, Father Pietro Secchi, in 1869. The term, which does not have such artificial connotations in its Italian form of *canali*, was used for a series of long, straight, dark markings stretching over large parts of the surface.

Schiaparelli never at any time suggested that the canals were of artificial origin, but he does appear to have believed that they were genuine natural waterways, and he named them accordingly. Most of them he named after real or mythical terrestrial rivers, such as Nilus, Euphrates and Lethe. Where two or more of these canals intersected, Schiaparelli mapped a dark circular area, which he may have believed to be an oasis, and he gave these some of the most delightfully euphonious place names in the solar system: Juventae Fons (Fountain of Youth), Lunae Lacus (Lake of the Moon), Niliacus Lacus (Egyptian Lake), Nodus Gordii (Gordian Knot).

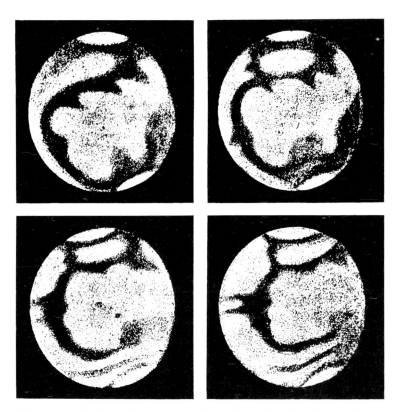

Figure 6.6.   Belief in the existence of 'canals' originated in drawings like these by Fr Secchi, made in Rome in 1858. These careful sketches have no suggestion of artificiality about them, and bear no relationship to the elaborate linear networks drawn by later astronomers.

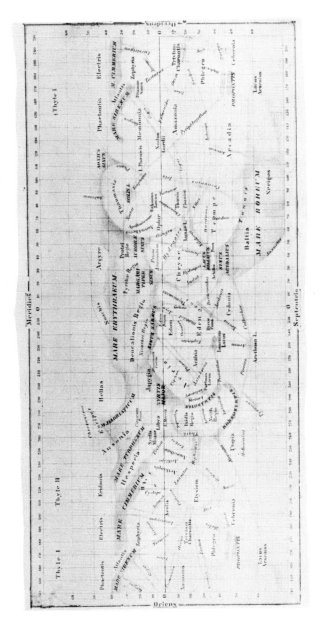

Figure 6.7. A detailed map of Mars, made by Schiaparelli in 1883. His 'canals' are much more numerous and linear than those drawn by Secchi, and all of them have grandiloquent mythical names.

Although Schiaparelli was cautious in his interpretations of the Martian canals, other astronomers were less so. One, Percival Lowell, a wealthy American amateur, became firmly convinced that Mars was populated by a race of intelligent beings, struggling desperately for survival in a desiccating climate by building enormous waterways to carry water away from the poles. So convinced was he, that he founded the Lowell Observatory at Flagstaff, Arizona, in 1894, specifically to observe Mars. Over the years, Lowell built the observatory up into the excellent institution it still is, and spent long periods studying Mars, constantly trying to improve the maps of the canal system. Eventually, his maps showed the whole surface of the planet covered in a cobweb of canals.

He wrote three books on Mars, in which he developed his extravagant ideas about the activities of Martians and their enormous civil-engineering enterprises, but even he was aware

Figure 6.8.   Percival Lowell, founder of the Lowell Observatory at Flagstaff, Arizona, a lifetime advocate of intelligent life on Mars and directly responsible for the discovery of the planet Pluto. (Photo courtesy of W. G. Hoyt and Lowell Observatory)

MARS—1905.

Figure 6.9. *Compare this map, drawn by Lowell in 1905, with Schiaparelli's (Figure 6.7 page 217). Many of the same features can be identified, and some of the 'canals' appear on both maps. Lowell's canals, however, are much narrower, and are often shown as pairs or 'twins'. (In comparing features, note that Schiaparelli's prime meridian is in the centre of his map, and Lowell's is at the far left.) (Lowell Observatory photograph)*

that no terrestrial telescope could hope to show anything so narrow as a canal on the surface of a planet tens of millions of kilometres away. Instead, he thought that the dark lines were belts of vegetation that grew up in response to irrigation water supplied by invisible canals.

Since it was impossible to *prove* Lowell wrong, and since many other astronomers went at least some way towards supporting him, belief in a fertile, inhabited Mars flourished until well into the 1960s, fertilized by a steady stream of books and endless tales of flying saucers and UFOs. H. G. Wells, the first great British writer of science fiction, and a trained scientist himself, was much influenced by Lowell, and wrote a classic book, *The War of the Worlds*, describing a Martian invasion of Earth. Decades after its publication, this book caused a panic in New York when it was broadcast as a radio play and misinterpreted by listeners who thought it was an account of a Martian invasion in progress.

Spacecraft pictures show nothing remotely like the canal systems that figured on so many astronomers' maps. So how on earth could so many observers have been so misled?

The psychology of perception seems to be entirely responsible. The human eye tends automatically to look for order in random patterns; to join up isolated dots into solid lines. And, of course, once the *idea* of straight lines has been seeded in the mind of the observer, he will never be able to eliminate it from his subconscious, no matter how hard he tries consciously to be objective. This may seem commonplace at first sight, but it is surprising just how much faith telescopic astronomers were prepared to put in their own visual observations. The American astronomer, Fred Whipple, writing in an earlier Pelican book, *Earth, Moon and Planets*, was himself prepared to write of an observer commencing to look at Mars: 'by persistent observation, night after night, [his] eye will become more and more expert until he is able to distinguish surface details that were at first completely invisible'. There may, indeed, be some improvement of visual acuity with experience, but it is now clear that most of the extra details that emerged on Mars came from the minds of the observers, not through their telescopes.

Phobos and Deimos: Fear and Fright

Good luck plays a part in many scientific discoveries. Johannes Kepler himself had more than his share of fortune in making his discoveries of the laws of planetary motion. But the discovery of the two moons of Mars was anticipated by what must surely be one of the most amazing pieces of guesswork in scientific history.

In *Gulliver's Travels* (1727), Dean Swift wrote of the astronomers of Laputa, the flying island:

They likewise have discovered two lesser stars, or satellites, which revolve about Mars, whereof the innermost is distant from the primary planet exactly three of his diameters, and the outermost five; the former revolves in the space of ten hours, and the latter in the space of twenty one and a half, so that the squares of their periodical times are very near in the same proportion with the cubes of their distances from the centre of Mars, which shows them to be governed by the same law of gravitation that influences the other heavenly bodies . . .

One hundred and fifty years later, in 1877, Asaph Hall, observing Mars through a telescope at the US Naval Observatory in Washington, discovered a faint object near Mars. After keeping watch on this for some time, he realized that the object was in orbit around Mars, and must be a satellite. A few days later, while looking out for this first moon, he discovered that there was a *second* one in attendance on Mars. Following the suggestion of an erudite Englishman at Eton, he named the satellites Phobos and Deimos, since these are the attendants of Mars referred to in a classical source, the fifteenth book of Homer's *Iliad*.

More recent work has shown that Phobos, the inner moon, orbits one and a half Mars diameters above the surface, and Deimos just over three. Phobos goes round Mars in 7·65 hours, and Deimos in 30·3. So Swift was startlingly close to the mark in his descriptions in *Gulliver's Travels*. But how did he do it? Was it just *pure* guesswork?

It now seems as though both logic and luck played a part; that Swift was right for the wrong reasons. He was a keen astronomer, and, more important, had read the works of Kepler carefully. Now, as we have seen already, Kepler had a large bee in his bonnet about regularity in the solar system. Although he could not reconcile his observations on the shapes of the orbits of the planets with the five perfect Platonic bodies, he was still convinced that the solar system was governed by rules and numbers. He knew that the Earth had one moon and Jupiter four, so he immediately thought that a neat mathematical doubling series existed: Earth 1; Mars 2; Jupiter 4; Saturn 8; and so on. (This clearly has something of the flavour of Bode's Law, although that had not as yet been discovered.)

Swift's remarkable predictions were clearly influenced by Kepler's ideas, but although they came so near to being right, there was no scientific basis for them whatever. There is some dubious theoretical data to support Bode's Law, but there is *no* simple rule controlling the distribution of satellites. Furthermore, Phobos and Deimos are both so tiny (neither is bigger than 30 kilometres in diameter) that they scarcely merit the title moons – they are more like odd scraps of inter-planetary junk picked up by Mars at some remote point in its history.

Some statistics

The reader may have gained the impression from the preceding accounts of the canals and moons of Mars that every astronomer who turned his telescope on to the planet was a slightly scatty individual whose mind never descended to earthly levels – the standard caricature of an astronomer. Nothing could be further from the truth. A great deal of solid scientific work was done on Mars by telescopic observers of the nineteenth and twentieth centuries, which paved the way for the later spacecraft missions. It would burden this account unduly to describe all this work adequately, but some of the information gained is of such importance that it needs to be noted here, because it provides a framework for understanding the discoveries of the space missions. First, the basic statistics.

The *mass* of Mars was quite easy to determine, since it has its two moons. Asaph Hall, discoverer of Phobos and Deimos, was the first to calculate the mass; he came up with a figure of $6·43 \times 10^{23}$ kg. This figure has been subsequently slightly refined, but the important thing is that it shows that Mars is a *light* planet, with only about one tenth of the mass of the Earth.

The *size* of Mars might seem at first to be even easier to measure, but some difficulties were encountered in doing this, because, when photographed in blue light, the planet appeared to be 3 per cent bigger than when it was photographed in red light. This surprising discrepancy probably originated in the atmosphere of Mars, although the details were not fully understood. It was learned, though, that the diameter of Mars at the equator is about 6,760 kilometres, and that, like the Earth, Mars has the shape of a slightly squashed orange, so that the polar diameter is about 40 kilometres shorter than the equatorial diameter.

The mass and size of the planet are not very interesting in themselves, but, taken together, they reveal the *density* of the planet, which is where geologists get interested. Mars has a bulk density of just under $4·0 \times 10^3$ kg m^{-3}, considerably less than the Earth's $5·5 \times 10^3$ kg m^{-3} value. So, it is immediately obvious that the internal structure of Mars must be different from the Earth. Since much of the mass of the Earth is concentrated in its metallic core, it is unlikely that Mars possesses such a massive core.

Early thoughts on the atmosphere

In the early days, it was natural to suppose that Mars had a thick, congenial atmosphere, capable of supporting Earth-like life. When the details of size, mass and surface temperature became known, and the all-important escape velocity could be determined, this rosy picture faded. Mars is a distinctly chilly planet, where surface temperatures are typically around -60 °C. This means that, although Mars is small, the escape velocity is relatively high and some gases can be retained. Carbon dioxide, one of the denser gases, was always regarded as a strong possibility, oxygen and water vapour rather less so.

A great deal of effort was put into trying to determine tele-scopically both the atmospheric pressure and the composition of Mars' atmosphere. Both are tricky determinations to make, the composition being particularly difficult owing to interference by gases within the Earth's own atmosphere.

Earth-based measurements of Mars' atmospheric pressure were made initially by investigating the scattering and polarization of light in its atmosphere. Early estimates erred consistently on the optimistic side, and as more and more refined measurements were made, the maximum atmospheric pressure had to be steadily whittled away. The first determination put the value as high as 100 millibars, one tenth that of the Earth's, but, later on, it shrank from 100 to 86, then to 50, then 20, then 15, and finally, when the spacecraft missions were being planned, to less than 10 millibars.

While it was still possible to talk in terms of fairly advanced life forms existing on Mars with an atmospheric pressure of 100 millibars, when the estimates were revised down to less than 10 millibars most scientists realized that the scales were becoming heavily loaded against the existence of advanced organisms.

Although attempts had been made since the 1880s to determine the composition of Mars' atmosphere, no significant progress was made until 1947, when G. H. Kuiper was able to confirm that carbon dioxide was present, and that it probably constituted a greater proportion of Mars' atmosphere than it did of the Earth's. Water vapour was also identified, and lent some support to the supposition that the gleaming polar caps were made of water ice rather than of carbon dioxide ice, as some astronomers had argued. More sophisticated measurements seemed to show that there were seasonal variations in the amounts of water in the Martian atmo-sphere, but all the evidence indicated that the total amount of water was small, the best estimates suggesting that if all the water in the planet's atmosphere were condensed, it would form a film only 40 microns thick (40×10^{-6} m). This is no more than enough to fill one average Scottish loch.

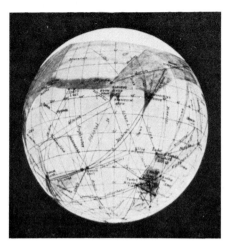

Figure 6.10. A globe of Mars, constructed by Lowell in 1905. Compare this with Figure 6.11! (Lowell Observatory photograph)

Figure 6.11. One of the best telescopic pictures of Mars ever obtained, this one was taken by R. B. Leighton with the Mount Wilson 60-inch telescope. Apart from the striking south polar ice-cap, the most important aspect of this feature is the lack of visible detail. The dark patch at the bottom right is the Syrtis Major, first recorded by Huygens in 1659.

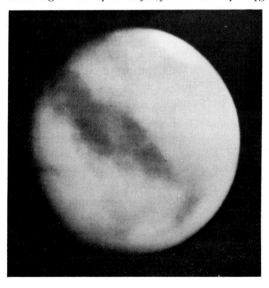

The combination of extremely low temperatures, a tenuous, carbon dioxide-rich atmosphere and the lack of water might seem sufficient to discourage *anyone* from believing in the existence of life on Mars. But this is to discount the powerful psychological forces at work. Percival Lowell saw canals on Mars basically because he *wanted* to see canals. Even in the twentieth century, there were many scientists who wanted to believe that life existed on Mars, and who therefore allowed themselves to dream up elaborate scientific rationalizations to justify their beliefs. The belief in the existence of life is in itself an exceedingly hardy organism; it has survived even into the post-Viking stage of Mars exploration, as we shall see. First, though, a brief discussion of the predecessors of the Vikings, the Mariner series of missions.

The Mariner missions

When the construction of space probes first became possible in the 1960s, it was natural that Mars would be high on the list of priorities. The first attempts were made by the Russians in 1962. Their Mars 1 spacecraft was launched successfully, but communications with it were lost when it was 106 million kilometres distant from the Earth. Their Zond 2 mission also failed in 1964.

The first Mariner mission to Mars was Mariner 4.★ It swung past Mars on 14 July 1965 on a 'fly-by' mission, passing only 10,000 kilometres above the surface and transmitting back twenty-two television pictures to Earth before drifting on into the oblivion of deep space. Since these were the first pictures of the surface of a planet which had captured men's imaginations for centuries, their arrival was awaited with tremendous interest and excitement. When they finally came, they proved to be something of a disappointment.

Fuzzy though they were, the pictures were good enough to dispel any lingering hopes that Mars might be a congenial place. There were no canals, no oases, no rivers, no forests; nothing to

★ Mariner 1 failed on launch; Mariner 2 successfully flew past Venus in 1962; Mariner 3 was targeted for Mars, but also failed on launch.

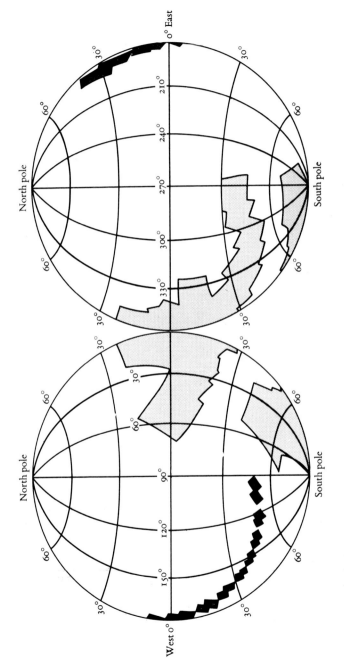

Figure 6.12. The areas of Mars covered by Mariner 4 photography (black) and Mariners 6 and 7 (shaded). Some rather misleading conclusions were drawn from this rather patchy coverage.

suggest the slightest signs of life. All the pictures revealed was a drab, cratered surface looking depressingly like the sterile wastes of the Moon and Mercury. Many scientists gave up any idea of Mars being an 'interesting' planet at that time. With twenty-two pictures in their hands, covering less than 1 per cent of the surface, they were inclined to write the whole planet off as being something like the Moon, only a bit bigger.

Two other Mariner 4 studies added weight to this rather gloomy view of the Red Planet. Studies of the refraction of radio waves as the spacecraft passed behind Mars confirmed positively that the surface atmospheric pressure is only 0·8 per cent of the Earth's (about 8 millibars), while studies on the magnetic field around Mars showed it to be extremely weak, less than 1/3,000th of the Earth's. This, as in the case of the Moon, suggests that Mars lacks a liquid metal core, and, by implication, probably also lacks the interesting dynamic processes that the Earth possesses.

Despite this rather disappointing start, several other missions in the Mariner series were planned to study Mars. Mariners 6 and 7 were also 'fly-by' missions, whereas Mariners 8 and 9 were intended to orbit the planet. (Mariner 5 went to Venus.) Mariners 6 and 7 both secured photographs, but they did little to dispel the rather dull impression created by Mariner 4.

Mariner 9

Mariner 9 was one of the most successful missions in the short but dramatic history of space exploration. Mars came particularly close to the Earth during 1971, so that year was an ideal one for exploring the planet. Two Russian and two Mariner spacecraft were launched. The Russian craft were intended to soft-land on the surface, and the Mariners to orbit the planet and obtain imagery of as much of it as possible, and also to examine the atmosphere.

Both Russian attempts were failures. The first seems to have crashed straight into the surface, digging a new crater for itself.

The second appears to have got down in one piece, but seems to have stopped transmitting data twenty seconds after touch-down; the landing was probably not as soft as it might have been. The first of the American pair, Mariner 8, also came to grief. It had a career that was as short and inglorious as Mariner 9's was long and heroic; something went wrong with the guidance equipment during launch, and it plunged into the Atlantic.

The Viking missions which followed Mariner 9 would not have been possible without the data obtained by this remarkable spacecraft; it is difficult to convey quite how enormously our knowledge of Mars was extended by Mariner 9. In some ways, its effects could be likened to handing Christopher Columbus *The*

Figure 6.13.   One of the first Mariner 9 pictures of Mars. Little surface detail can be seen, since a great dust storm had drawn an opaque veil over most of the planet. The three dark spots are major volcanoes, rearing above the dust clouds.

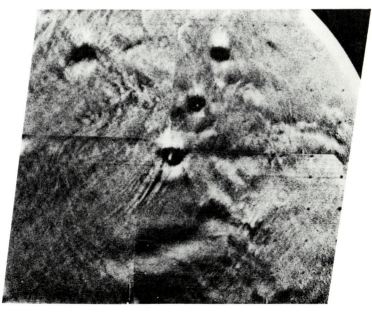

Times Atlas of the World before he set off to cross the uncharted expanse of the Atlantic. Even this would be an understatement, though, because Mariner 9 sent more than 7,000 images back to Earth, and relayed a total of more than 30 billion 'bits' of information, an amount equivalent to more than thirty-six times the entire text of *Encyclopaedia Britannica*.

The first pictures from Mariner 9 were also something of a disappointment. A planet-wide dust storm was raging when it arrived, and all that the pictures showed was rolling yellow clouds obscuring the whole surface. As the weeks passed, the storm abated, however, and as soon as the Martian skies had cleared sufficiently, a steady stream of pictures began to reach Earth.

The first task that the Mariner 9 science teams were faced with when they received the great flood of pictures was to use them to build up a topographic map of the planet. Only a few features on the photographs could be correlated with those shown on the maps made by astronomers peering through Earth-based telescopes, so they effectively had to start mapping from scratch. Perhaps the least of their problems was that of nomenclature for the hundreds of craters, volcanoes, valleys and plains that they found. In many cases, they fell back on the elaborately elegant Latin names given by Schiaparelli and Lowell to areas which they identified as the sites of 'canals' and 'oases'. As a consequence, even the most modern maps of Mars are strewn with impressive polysyllabic names, with a thoroughly antique flavour.

The sheer volume of Mariner 9 data is something of an embarrassment, particularly to those who would make so bold as to try and describe a complete planet in a single chapter. Imagine trying to describe the geography of the whole world in a few thousand words. What would you leave out? If you attempted to describe rivers and glaciers in a meaningful fashion, would you also be able to include coral reefs and sandbanks, as well as Siberian tundra and African rift valleys, and the Nullabor Plain and continental shelves and the isthmus of Panama and the Mid-Atlantic ridge and island arcs and the Himalayas . . . ? Clearly, one has to stick to generalities.

Mariner 9 and the geology of Mars

Once the basic topographic maps were available, geologists could set to work, constructing the first geological maps and trying to work out the main events in the history of the planet as well as how its structure differs from that of the Earth. A most important discovery was that, while there are no 'continents' or 'oceans' on Mars – this came as no surprise – there is a major contrast between the northern and southern hemispheres. The southern hemisphere consists of heavily cratered terrain, which shows all the signs of having been severely battered during a period of intense bombardment very early on in time, probably more than 4 billion years ago. In this respect, the southern hemisphere of Mars is strikingly similar to the lunar highlands. There are even large circular impact basins on Mars, such as Argyre and Hellas, which are analogues of the great lunar basins, Imbrium and Orientale.

The northern hemisphere is quite different. It consists mainly of vast expanses of relatively crater-free 'smooth' plains, somewhat similar to the lunar maria. Like the maria, the Martian plains are covered by basaltic lava flows. Like the maria, too, these lava plains are peppered with small impact craters and are traversed by long, narrow, winding wrinkle ridges.

The radiometric dating of samples from the lunar maria showed unambiguously that those lavas were erupted between 3·9 and 3·2 billion years ago. Since there are no samples yet available from Mars, there is no way that the ages of Martian rocks can be determined directly. It is possible, however, to use statistical analyses of crater size and frequency distributions to estimate the ages of surfaces on the planet. (These techniques were discussed briefly in Chapter 3, pages 94–5.) Such methods rest heavily on the assumption that Mars' cratering history has been closely similar to the Moon's, though this is rather a bold assumption!

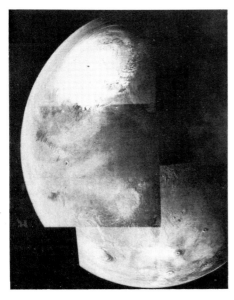

Figure 6.14.   A Mariner 9 view of Mars, showing the northern hemisphere after the dust storm had abated. Olympus Mons and three other great volcanoes are visible in the lower right. All that can be seen in the rest of the picture are the vast, uniform expanses of the northern hemisphere smooth plains.

Two results emerge from crater density studies which are fundamental to the history of Mars. The first is that the vast Martian smooth plains are of broadly the same age as the lunar maria, and the second is that the lack of small craters on Mars compared with the Moon indicates that there must have been a major episode of erosion which smoothed out most of the earlier formed small craters. This episode may have taken place as much as 2 billion years ago, though the figure should be treated with caution: crater statistics on Mars are constantly being revised, and many alternative ages for different features have been proposed.

In possessing cratered highlands and lava plains of great antiquity, Mars is very like the Moon. But here the similarity ends. After the eruption of the vast basaltic lavas on the Moon, little has happened there, except for a few feeble moonquakes and the occasional meteorite impact: the Moon is dead. But

Mars is a living planet. Great volcanoes have been constructed, and may still be active; rift valleys have opened, and may be opening still; great rivers once flowed, and hurricane-force winds still scour the surface, shaping and sculpting it.

Volcanoes and plate tectonics

In terms of sheer physical size, the volcanoes of Mars are its most impressive features. While the vast dust storm was raging when Mariner 9 first entered Mars' orbit, the only features that could be distinguished were four great volcanoes which reared their heads high up above the swirling dust clouds. One of these could be identified with a whitish spot shown on maps made by telescopic observers, and named by them Nix Olympica, the Snows of Olympus. The others were entirely unknown, and were initially given rather prosaic names: North, Middle and South Spots. Later, when more time was available for such refinements, they were christened Ascreus Mons, Pavonis Mons and Arsia Mons respectively. (During this spurt of enthusiasm for name-giving, by the way, Nix Olympica lost its snows, and became merely Olympus Mons.)

The four volcanoes are grouped together on an elevated bulge on the surface of Mars known as the Tharsis Dome region, which is itself 7 kilometres higher than the surrounding regions. Olympus Mons is much the biggest of the four. Although it has a profile which is similar to that of 'shield' volcanoes on Earth (like those of Hawaii), the shield of Olympus Mons is more than 500 kilometres across, while its summit is no less than 23 kilometres above the surrounding plains. By contrast, the largest volcano on Earth, Mauna Loa in Hawaii, is only 200 kilometres in diameter, and rises a mere 9 kilometres above the Pacific Ocean floor. True to its name, Olympus Mons is a fitting home of the gods, and to make their abode more impressive still, the whole edifice is ringed by a formidable sheer cliff, rising 4 kilometres above the plains.

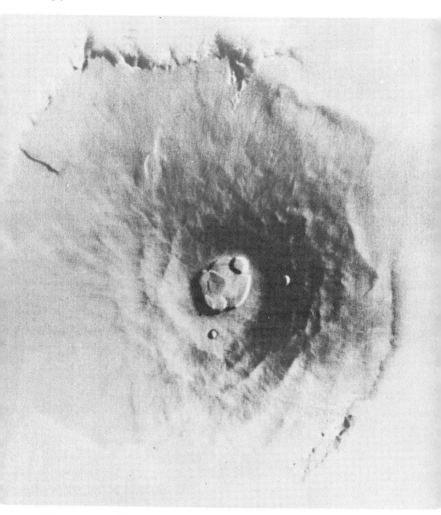

*Figure 6.15.   Olympus Mons, the largest volcano in the solar system.
Note the complex set of nested craters at the summit, and the ring of cliffs
around the base. These cliffs are up to 4 kilometres high.*

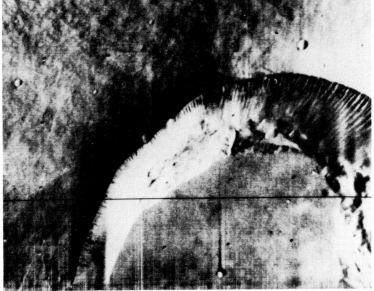

Figure 6.16. *The summit crater of Pavonis Mons, one of the major volcanoes seen in Figure 6.14 above with (at right) a close-up of the crater wall. Some impact craters can be seen on the flanks of the volcano, giving a guide to its age. The fluting of the crater wall is probably the result of wind erosion, while the smooth crater floor may represent an old lava lake.*

If one excludes, for the time being, the huge but doubtful features on Venus, Olympus Mons is by far the biggest volcano in the solar system. Many other volcanoes exist on Mars, and they also are enormous. Why? The answer must lie in the internal structure of the planet. The crust of the Earth is quite thin – only about 10 kilometres in oceanic areas – and its interior is hot and mobile. Consequently, the crust is prone to break up and shift around on the surface, adjusting itself to changes taking place in the interior. These movements are known collectively as *plate tectonics*, and they lead to the development of chains of volcanoes in distinctive environments along the margins of the great crustal plates, such as the circum-Pacific 'Ring of Fire'. One of the consequences of plate tectonic movements, however, is that no terrestrial volcano ever stays in one place for any length of time; it is carried away on its crustal plate from the 'hot spot' in the mantle that gave it birth.

On Mars, there is no evidence that plate tectonics has ever operated. The nearest approximation to it is a great rift, Vallis Marineris, which slashes for 5,000 kilometres across the centre of the planet, one end of it being located in the Tharsis Dome region. This rift may mark a site where continental drift attempted to get under way but failed. In some respects, the combination of up-doming and rifting displayed on Mars is akin to the early stages of the formation of rift oceans such as the Red Sea on Earth. The reason why the Martian rift was stillborn, however, probably lies in the fact that the crust of Mars is very thick indeed, possibly as much as 200 kilometres. Such a crust would be immensely strong and rigid, and it would not be capable of breaking up into large, freely moving pieces. Thus it appears that Martian volcanoes are bound to remain fixed over their internal hot spots, perhaps for hundreds of millions of years. It follows that, even if the rate of eruption of Olympus Mons was as little as 0.1 km^3 of lava per year (the same rate as that of the Hawaiian volcanoes), the cumulative effect would be that the volcano kept on and on growing, leading to the huge structure that exists today.

One difficulty in this kind of hypothesizing about the giant Martian volcanoes is that it is impossible to be certain just how

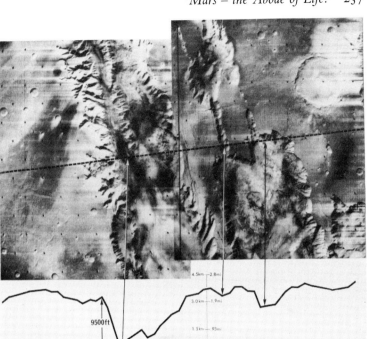

Figure 6.17. Part of the enormous Vallis Marineris canyon complex, with a topographic profile. The Grand Canyon of the United States is about the same size as one of the small tributaries on the left of the main canyon.

old they are, and for how long they have been active. Some of the early studies of the Mariner 9 photographs suggested that they might be quite young, and, indeed, that some might be still active. But, as more and more data accumulates, particularly the new generation of high-resolution Viking pictures, it is beginning to look increasingly as though most of the volcanic history of Mars took place early on in its evolution, as did that of the Moon. Thus, while some early estimates of the age of Olympus Mons were around 200 million years, more recent estimates are several orders of magnitude older, around 2·5 *billion* years. This is not to suggest that *all* Martian volcanoes are exceedingly old. Some of the superb Viking photographs show apparently young, fresh-looking volcanic features, so that there is still a chance that some lingering volcanic activity continues at the present day.

The ice-caps

Ever since Cassini first glimpsed the gleaming white polar caps on Mars in the seventeenth century, they have promoted specu-lation about their influence on conditions on Mars. Mariner 9 and the two Viking orbiters provided some extremely important new data on the ice-caps, but even now their influence is not fully understood.

The Mariner 9 pictures revealed that the ice-caps are surrounded by complex mantles of wind-blown dust forming flat-lying layers which have themselves been sculpted by the wind, so that the terrain has a finely laminated appearance. Near the north pole, there are huge expanses covered by dunes. These may have been formed by winds blowing over the laminated terrain; the smallest, dust-sized particles would be blown clear away, but larger-sized grains would be swept across the surface, and would accumulate in dunes.

One of the key areas of debate concerning the ice-caps has always been over exactly what the 'ice' is. Ordinary water ice and carbon dioxide ice ('dry ice') as well as more complex mix-tures have all been proposed. It is often said that the weather that we experience – or suffer – on Earth originates in our polar regions. The Antarctic ice-cap is particularly important, especially since it contains the majority of all the fresh water on Earth. If our ice-caps melted, the Earth's climate would be drastically different.

On Mars, the ice-caps are even more important. At one time, it was widely thought that the ice-caps were made of carbon dioxide ice. This led to suggestions that, if the temperature on Mars were to rise slightly, the polar ice would melt and the atmospheric pressure increase. The increased carbon dioxide

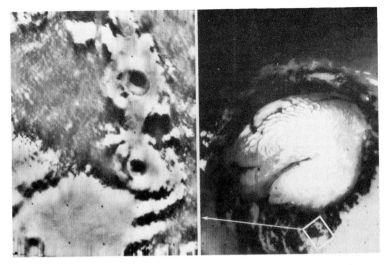

Figure 6.18. *The north polar ice-cap (right) and a close-up view of part of the dark mantle that surrounds it. Changes in this mantle had been observed telescopically and were used to argue for the existence of seasonal changes in a hypothetical vegetation cover.*

Figure 6.19. *The layered terrain in this view near the south pole of Mars may conceal massive amounts of water. The layers may consist of dust or volcanic ash containing water or, possibly, carbon dioxide ice. They overlie older, hummocky terrain at the bottom of the picture.*

content of the atmosphere would then yield a greenhouse effect, and the temperature and pressure would increase further until a pressure approaching the surface pressure on Earth was reached. In this way, it was argued, Mars could at some periods in its history have been a far more hospitable place than it is now.

Spacecraft data have damped down these rather cosy conjectures to some extent. The polar caps are now known to consist of *two* distinct components: one ephemeral, the other permanent. During the Martian winter, the polar ice advances nearly as far as 30° of latitude. This ice is 'dry ice', frozen carbon dioxide, and generally forms only a thin layer, rather like a heavy hoar frost, though several metres' thickness may accumulate in the polar regions. With the advance of the Martian summer, the dry ice sublimes, passing straight from solid to gas, and the ice-cap rapidly shrinks. Even in the height of summer, however, a residual cap remains. Because the polar regions are much warmer in summer (at a chilly -68 °C) than would be possible for the existence of carbon dioxide ice, the permanent ice-cap is thought to consist of ordinary water ice. Vast amounts of water may be locked up in the polar caps, but it is impossible to be certain how much, and therefore what part it may have played in the geological evolution of Mars.

Dust to dust

One of the factors which fertilized beliefs that there might be life on Mars was that astronomers could see seasonal changes taking place in the appearance of the planet. Most conspicuous of these were the successive advances and retreats of the polar ice-caps, but apparently linked to them were the less obvious changes in the dusky markings at lower latitudes. These latter were identified as being due to seasonal variations in vegetation, as the changes in the ice-caps alternately made more or less water available. All this, of course, was grist to the mill for the science-fiction writers, who made great use of it in their stories of an intelligent humanoid race struggling to support existence on a progressively desiccating planet. Sadly, the SF stories were way off the mark, but the true situation is almost as fascinating.

The dusky markings are not, of course, anything to do with vegetation. Sophisticated telescopic studies showed that the darker markings were merely a darker red than their surroundings, and that this colour difference was probably caused by a difference in the oxidation state of the surface material – some of it was a rustier red than the rest. There are, indeed, 'seasonal' changes, but these are the result of Martian winds which scour across the surface, lifting dust in some places and redepositing it in others. Over the months of the Mariner 9 mission, individual blotchy markings could be seen to change in shape from one picture to the next, as light-coloured dust was blown off areas with darker backgrounds.

Dust storms were well known to take place on Mars prior to the Mariner 9 mission, but Mariner 9 emphasized how severe these dust storms are, and how important winds are as agents of erosion on the planet. It is now known that surface winds may be as fast as several hundred kilometres an hour. Winds of that speed on Earth would be extraordinarily damaging, but because the Martian atmosphere is so tenuous, the situation is quite different. Speeds of about 200 kilometres an hour are required to lift dust particles off the Martian desert surface. This means that dust particles are not often lifted, but when they are, they travel at such high speeds that they are many times more effective as scouring agents than they would be on Earth; and so, in effect, the dust storms act like sandblasting guns, carving into mountains, volcanoes and rocks.

Much of the surface of Mars shows evidence of this sculpting by wind. Some large impact craters have long chains of dunes on their floor, strikingly similar to the wandering groups of dunes in the Sahara Desert; other craters have had their profiles softened and eroded away until they are barely recognizable. The great barrier of cliffs that rings Olympus Mons may itself have been carved by the abrasive Martian winds.

Figure 6.20. Light and dark blotches on the surface of Mars. Mariner 9 scientists were able to track changes in the shape of individual blotches during the course of the mission, in particular of the leaf-shaped dark patch at the centre of this photo. Similar, larger-scale, changes had been used in the past to argue for vegetational growth. The changes are undoubtedly caused by wind erosion lifting and redepositing dust.

Canyons and channels

Wind is a much less important agent of erosion on Earth than is water, except in a few particularly arid areas. On Mars, the situation is reversed: water-eroded land forms are much less abundant than wind-carved ones. It is remarkable enough that such features exist at all on a planet which lacks oceans, rivers, lakes or even ponds of any description. It is ironic, too, that the spacecraft which finally dispelled any vestige of hope that there might be 'canals' on Mars actually demonstrated the presence of quite different water-eroded channels.

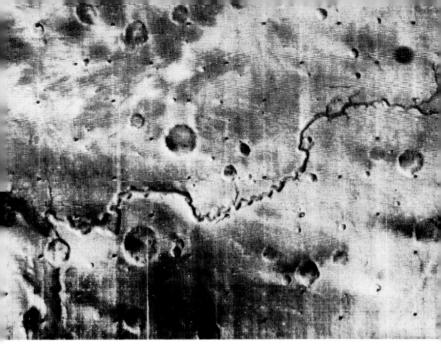

Figure 6.21. One of the most unexpected discoveries made by Mariner 9: a sinuous channel, some 40 kilometres wide and 5 kilometres deep. The meandering course and tributaries feeding this channel strongly suggest an origin by flowing water.

These remarkable features were undoubtedly the single greatest discovery made by Mariner 9. Three different types seem to exist. First, there are myriads of small, dendritic channels which drain downwards from elevated regions. These are interpreted as simple 'run-off' channels, the sort that would be expected to form after heavy rain in a dry, hilly area. Next, there are much bigger, broader channels cut into the cratered highland areas of Mars which wind downslope, debouching on to the smooth plains of the northern hemisphere. These often have minor tributaries feeding into the main channel, and show braided patterns like those formed in terrestrial river channels by flash floods. Finally, there are channels which, when traced upstream, end abruptly in cliff-like features which show clear evidence of landslipping on a massive scale.

Figure 6.22.   Another great channel, winding through the ancient cratered highlands of Mars. The channel is clearly younger than almost all craters, but may still be very ancient in terrestrial terms.

Figure 6.23.   A close-up Mariner 9 view of one of the channels. Flow was from left to right. Note the complex branching and braiding of the channel at right. This is common in terrestrial river channels in desert environments.

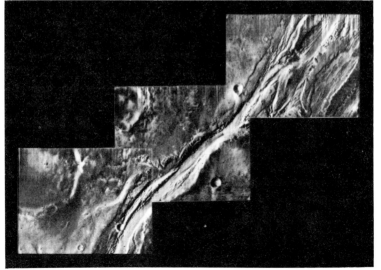

It seems inescapable that the Martian channels were formed by water erosion; that great rivers once flowed through the sterile deserts. But there is *no* liquid water now, and none could exist on the surface because of the low atmospheric pressure. If a stream of water were exposed, it would first begin to evaporate rapidly, and then evaporative cooling would cause a film of ice to build up over the surface, reducing the rate of evaporation, but also preventing easy flow.

A close examination shows that the Martian channels are by no means identical to terrestrial rivers, and that some different mechanisms must have operated in their formation. For example, the meandering Martian channels are in general much wider for a given size of meander than terrestrial ones. One possible process that could have enlarged the channels is *artesian sapping*. On Earth, some river valleys are broader than could be expected from the size of the river flowing through them – an example is the Grand Canyon. Once the valley has been cut, ground water seeping slowly out of the rocks exposed in the walls saps away at them, undermining them and eating back until eventually the width of the channel is considerably increased. The same could have happened on Mars.

But how and when could *any* significant volumes of water have flowed on Mars? This is the single most important question for students of Mars. *If* liquid water once flowed in the great channels, *life* might have been able to evolve in the relatively hospitable environments along the banks of the channels.

Crater density studies reveal one important clue: there was no *single* episode of channel formation on Mars. The oldest channels are probably as much as 4 billion years old, the youngest could be as little as 500 million years old. The present Martian climate is an extremely arid, severe one. But the mere existence of the channels tempts one to think that it must have been better and wetter when the channels were cut. One does not need to look far to find evidence of major changes in planetary climates with time – our own Earth is just emerging from (or may still be experiencing!) a great Ice Age, and it has suffered two others in the last 600 million years. One of the factors which seems to have triggered the Earth's ice ages are subtle changes in its orbital characteristics. Mars may have experienced similar but more profound changes.

Mars' planetary neighbour is the giant planet Jupiter. Now, although Jupiter is always hundreds of millions of kilometres from Mars, its mass is so great that, as the two planets orbit around the Sun, Jupiter perturbs the orbit of Mars from a nearly circular shape to a much more elliptical one. The change from circular to elliptical is a recurrent process, each cycle taking about 2 million years to complete. Its effect is that the mean distance from Mars to the Sun varies significantly over this period, and thus, during the periods when it is closer to the Sun, the climate is warmest.

Another source of climatic variation arises from the fact that Mars' spin axis is inclined to the plane of its orbit at an angle of 24·9°, closely similar to the Earth's. Although the angle itself remains steady, the effect of the Sun's gravitational pull on Mars' equatorial bulges is such that the spin axis does not always point in the same direction; it wanders or *precesses*, describing a small circle in space. The axis of a spinning top behaves in the same way as it wobbles uncertainly in the last few moments before toppling. Mars' axis takes about 100,000 years to complete one small circle, or wobble. This motion, superimposed on the changes in shape of the orbit, could cause significant and regular changes in the planet's climate.

It seems, then, that there are at least two ways in which the climate of Mars could vary quite dramatically. But the problem is by no means solved. Most geologists feel that the time scales over which the variations described operate are far too short to be responsible for the present surface features, all of which are extremely ancient by terrestrial standards. Furthermore, the likelihood that Mars' ice-caps are largely water has important implications for any ideas on large changes in climatic conditions. Even if changes in Mars' axial inclination or the shape of its orbit could cause changes in surface temperature, it is unlikely that these would have *major* effects on surface conditions. For liquid water to exist at the surface, a temperature increase of something like 70 °C would be required, according to some authorities, and this is most unlikely to be achieved as a result of any orbital changes. Thus, it seems that the polar ice-caps must be ancient features, and that no *major* climatic or atmospheric changes have taken place since they formed. Inevitably, however, conflicting views exist on this subject.

An alternative line of argument is that, since the great channels seem to have been formed by massive, short-lived deluges rather than by slow, steady flow, and since these deluges seem to have occurred sporadically over billions of years, then an *internal* origin is more likely than any calling for major climatic changes. The most plausible mechanism so far proposed is that great floods were triggered off by the sudden melting of large volumes of permafrost ice caused by geothermal heating. This would certainly produce large volumes of water very quickly. Examples are known on Earth, where volcanic eruptions have broken out beneath glaciers and ice sheets, particularly in Iceland. The massive melting caused by these glacier bursts or *jokúllhlaups*, as the Icelanders call them, liberates staggering floods of water which rival the Amazon in their rate of flow, and which are capable of changing the topography of large areas. Similar events taking place on Mars could account rather well for some of the great channels, except, unfortunately, for the fact that no obvious links can be seen between the source regions of the channels and any sites of volcanic activity.

Mariner 9: an epilogue

It is difficult to over-emphasize the achievement of the Mariner 9 mission. Not only did it provide a wholly new view of Mars, revealing it to be a planet with a rich and diverse geological history, but the pictures of Phobos and Deimos that it obtained were also the first ever obtained of *any* of the scores of minor bodies in the solar system. Mariner 9 spent 516 working days in space, and nearly a year in orbit around Mars. It transmitted 7,300 pictures back to Earth. On 27 October 1972, it used up the last of its attitude control gas and tumbled out of control, unable any

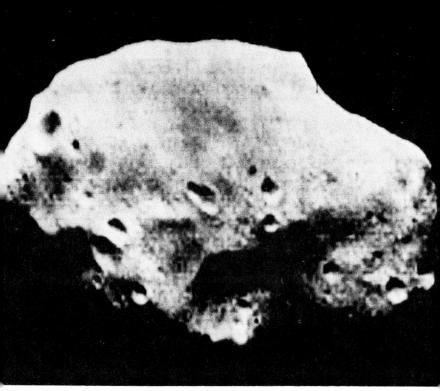

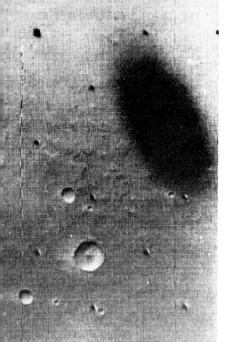

Figure 6.24. *The first-ever close-up view of Phobos, Mars' largest satellite, this picture was taken by Mariner 9 from a distance of about 5,540 kilometres during its thirty-fourth orbit of Mars. The satellite is strikingly irregular in shape, and is heavily cratered.*

Figure 6.25. *Another extraordinary Mariner 9 photograph: the shadow of Phobos on the surface of Mars. From the surface, Phobos would, of course, produce a total eclipse of the Sun.*

longer to keep its solar panels orientated towards their only source of energy.

Mariner 9 far exceeded the wildest hopes of the mission planners. Long before it died its distant death, however, another mission had already been planned, one that would be able to take the next logical step forward in the exploration of Mars: a mission to land on the surface of the planet and search directly for signs of life.

7. Project Viking

The original Vikings were a blond, blood-thirsty Nordic tribe that ranged around in longships, pillaging Christian settlements and burning everything they could lay their hands on. The new Vikings could not have been more different. The most sophisticated spacecraft ever launched, they were crammed with electronic equipment that had taken years to perfect and which represented the pinnacle of human achievement in technology. But they did have one thing in common with their predecessors: their missions to Mars were impelled by the same burning desire for discovery and exploration that drove the first Vikings to cross the Atlantic and set foot on the New World.

The twin Viking missions were initiated in 1968, well before Mariner 9 had made its epoch-making flight. Both spacecraft were based on the tried and tested Mariner design, but with many modifications. The most important of these was that each craft consisted of two modules, an orbiter and a lander. The two were united for the long space flight between Earth and Mars, and remained united initially in orbit around Mars. The cameras on the orbiter components then carefully examined the landing sites pre-selected from the earlier Mariner 9 photographs, to verify the suitability of each site and to search out more promising ones. Only when a safe site had been found was the decision made in each case to separate the landers from the orbiters, and to commit the landers to a descent. This represented a great step forward on the previous unsuccessful Russian attempts to soft-land on Mars. These were shots in the dark; once the spacecraft had been launched from Earth, there was little that could be done to guide them down to a safe landing. The Russians were trying to score a bull's-eye by throwing darts at a board so far away that it was almost out of sight; the Americans practically walked up to the board and stuck the darts in by hand.

Viking 1 was launched on 20 August 1975. After cruising through space for ten months along a smooth-curving trajectory, it reached Mars and went into orbit on 19 June 1976. The original mission plan called for the spacecraft to spend a fortnight orbiting the planet, scrutinizing the surface minutely in the 'site certification' process before landing on 4 July, the bicentennial of the American Revolution. This magnificent celebration of a great occasion was not to be, however. Viking 1's pictures of the 'prime' site selected for it showed that it was not nearly as smooth as it had seemed on the Mariner 9 pictures. On top of this, precise measurements by Earth-based radar equipment also suggested that the surface was too rough. Prudently, the mission controllers decided to delay the landing, and the lander craft remained safely in orbit while the bicentennial came and went.

Figure 7.1.   A model of the Viking 1 spacecraft in its preliminary orbital configuration. The broad panels are the solar cells for power generation; the lander is carried within the hemispherical shell.

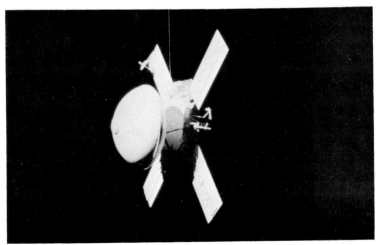

This was naturally a disappointment, but the delay itself was a tribute to the flexibility programmed into the Viking missions. Far better to wait a while than to commit a multi-million dollar spacecraft to an uncertain landing. As it turned out, the orbiter/lander combination remained together until 20 July before it was decided to attempt a landing in an area to the north-west of the original site.

The orbiting spacecraft were swinging around Mars at a velocity of about 4 kilometres per second. A carefully pre-programmed series of events was required to decelerate the lander from orbital speed and ensure that it touched the surface as gently as thistledown (nearly) at a speed of less than 3 metres per second.

Figure 7.2.   A diagrammatic summary of the complex sequence of events required to deliver the Viking landers safely to the surface of Mars.

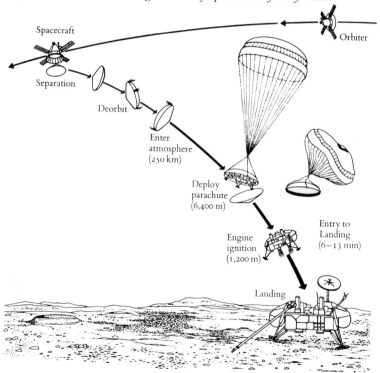

First, explosive bolts were detonated to separate the orbiter from the lander, and springs were released to push the two apart. Ten minutes later, the lander's rockets were fired to slow it slightly and hold it in the correct orientation for descent into Mars' atmosphere. Two or three hours after separation, the lander began to encounter the first wafts of the atmosphere, at a height of about 250 kilometres.

The lander was protected by a specially designed aerodynamic shield or 'aeroshell', which was able to maintain a stable attitude during the descent through the atmosphere, while simultaneously slowing the vehicle and dissipating the considerable heat generated by the passage through Mars' tenuous atmosphere. Peak deceleration occurred at a height between 30 and 24 kilometres above the surface. By the time it had reached 6 kilometres above the surface, the lander was travelling at only about 375 metres per second. At about 6 kilometres, a parachute popped out automatically to slow the rate of descent still further in the thicker part of the atmosphere, and the aeroshell, now redundant, was jettisoned. The lander swung on its parachute for only 45 seconds, until it reached an altitude of 1,200 metres and a speed of only 60 metres per second; at that point, the parachute was cut loose and three terminal descent engines fired briefly to bring the lander down to the surface. (The low atmospheric pressure on Mars meant that the parachute could not slow the lander sufficiently by itself, as it could have done on Earth.) The whole process from atmospheric entry to touchdown took only about ten minutes. In this short space of time, all of man's knowledge of aerodynamics and the Martian atmosphere were severely tested; no one could be sure that all the calculations would be correct. As it turned out, everything worked perfectly.

At the moment of the Viking 1 touchdown, Mars was 340 million kilometres distant from the Earth, so that the mission controllers in Pasadena, California, had to wait a nail-biting nineteen minutes from the time that they knew the landing *should* have taken place until the electronic signals, flashing at the speed of light between Mars and Earth, arrived to confirm that it had. At 5.12 Pacific Standard Time on 20 July 1976, the prosaic codeword ENABLET appeared on a television monitor at Pasadena, California. This meant that the three terminal descent engines had shut down, and that Viking 1 was sitting on the surface of Mars in one piece. Minutes later, the first crisp and detailed pictures from the lander cameras were being studied by the jubilant mission controllers. Coincidentally, the landing took place seven years to the day after Apollo 11 touched down on the surface of the Moon.

A few weeks later, on 2 September, the same brilliant team were cheering again, when Viking 2 also made a perfect landing and began to send a stream of pictures down to Earth. The team went right on cheering, too, because while the 'nominal mission' plans were based on everything coming to an end in December 1976, when Mars went behind the Sun, all systems were still working perfectly on both spacecraft when Mars re-emerged, and a so-called 'extended mission' was begun, which lasted until well into 1979.

Figure 7.3.   An artist's impression of the various Viking components. The lander is depicted securely on the surface, while the parachute attached to the jettisoned aeroshell has just touched down. (Not to scale.)

With four live spacecraft on Mars or in orbit around it, the volume of information reaching Earth during 1977 was phenomenal. It will take years to process all of it. Lengthy papers have already been published in the scientific journals on the achievements of Viking. Only a few of the highlights are picked out here.

The view from the orbiters

Although Mariner 9 succeeded in photographing almost the whole of Mars, most of its pictures were of relatively low resolution; they did not show features much less than 1 kilometre across. The Viking orbiter cameras were of basically similar design, but were able to cover a large part of the surface at higher resolution, at best showing features less than 50 metres across. Furthermore, they were also able to provide colour and stereoscopic images. With all this sophisticated imaging equipment at their disposal, the Viking team had almost as good a view of Mars as the superb LANDSAT satellites have provided of the Earth. They were thus capable of making better maps of Mars than existed of many underdeveloped countries on Earth, before LANDSAT appeared on the scene.

Figure 7.4.   An oblique view from the Viking 1 orbiter over Argyre Planitia, one of Mars' largest circular impact basins. Note the proliferation of channels around the small craters in the foreground. These are thought to have been cut by run-off of surface waters. On the horizon, some haze or cloud layers are conspicuous in the tenuous atmosphere.

Craters

When the various Mariner missions sent back their photos, it was generally thought that the Martian craters were similar to those of the Moon and Mercury. So similar, in fact, did they appear that no one was particularly interested in them. A few diligent scientists studied them closely, and found minor differences in morphology, but these were only to be expected, since erosion on Mars has softened the outlines of many craters, and the higher surface gravity on Mars means that ejected material was bound to be thrown less far than on the Moon. It was also correctly predicted that there would be far fewer *small* craters on Mars than the Moon – the tenuous atmosphere is sufficient to burn up most small meteorites.

The improved resolution of the Viking images revealed one remarkable difference in the ejecta blankets surrounding the craters. On the Moon, the impact of a meteorite on the surface hurls a blanket of ejecta upwards and outwards away from the impact site. The material rains back down around the impact site, producing a characteristic pattern of secondary craters. On Mars, by contrast, many craters look as if they were produced by meteorites splatting into a soft, wet surface. Most of the ejecta, instead of being flung upwards, seems to have flowed out radially in sludge-like masses. This entirely unexpected discovery at first caused some head-scratching, but it soon became apparent that there might be a simple explanation for the phenomenon.

When the impacts took place, the surface layers of Mars probably contained large volumes of water, possibly in liquid form, or more likely locked up as ice in the form of permafrost. At the instant of impact, the energy released vaporized this and caused radial flows of pulverized material, a bit like the base-surge deposits that are flung out by volcanic eruptions taking place in wet environments. On the Moon, of course, there is not, and almost certainly never has been, any water, so there the dry surface material was hurled up ballistically.

Figure 7.5. *Arandas crater, 25 kilometres in diameter, has a pronounced central peak, like many lunar craters, but displays spectacular 'ramparts' of material which appear to have splayed out radially in a sludge-like fashion. The dark band at the top of the crater wall suggests that two or more distinct layers of rock are present.*

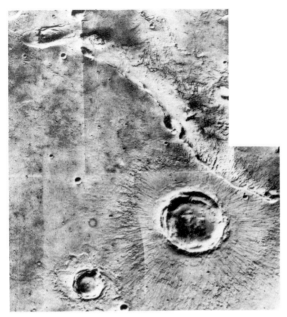

Figure 7.6.   The large crater in this photograph is 30 kilometres in diameter. It has a much more strongly radial pattern of ejecta than Arandas, but probably formed in a similar way. The ejecta spread out over the line of cliffs, demonstrating that the cliffs must have been eroded before the impact event. The ejecta also post-date the smaller crater at bottom left, which is similar to Arandas.

Glacial features

Although the highly distinctive craters are a good guide to the important part that permafrost has played in shaping the surface of Mars, they are by no means the only evidence. In arctic areas of the Earth where permafrost is present, it is well known that the ice-bound earth and soil resting on even rather gentle slopes tend to flow steadily downhill. Large volumes of material can be transported in this way, and severe problems can arise if houses or roads are built on these slowly sliding materials. This phenomenon is called, appropriately enough, *gelifluction*. In some areas of Mars, particularly near the boundary between the old cratered terrain and the younger smooth plains, the Viking pictures revealed many examples of isolated hills surrounded by aprons of debris, and of valleys along which debris seems to have flowed by a process similar to gelifluction.

Another characteristic feature of permafrost regions of the Earth is large expanses of 'patterned ground', in which the alternate freezing and thawing of water in the soil have caused polygonal ridges of stones to be built up at the surface. Similar features were discovered on the huge expanses of the smooth plains that were designated as the landing site for Viking 1. The plains are broken up into enormous polygonal fracture patterns, like those that appear at the bottom of a muddy pool that has dried out. Although they look similar to the polygons in terrestrial 'patterned ground', the Martian ones are several times bigger, reaching between 10 and 20 kilometres across. Since there are ice-caps on Mars, it is not perhaps surprising to find evidence of permafrost there. It is more surprising that permafrost features are so widespread on the planet. They leave little doubt that Mars must once have contained a great deal of water in the form of sub-surface ice.

Floods and channels

The lava plains of the Chryse Planitia area where the polygonal fractures occur were among the areas mapped by the early astronomers as being Martian 'seas' and 'oceans'. The Viking pictures reveal that these plains are now flat, dry wastes. It is difficult to imagine anything less like terrestrial seas. Oddly enough, though, the Viking pictures also show quite unambiguously that the same plains were once deluged by great floods of water sweeping across them.

The floods appear to have carved away an older series of strata which once covered the plains, leaving only occasional remnants sticking up as plateaux and mesas. In one area of Chryse Planitia, the plateau-forming materials have been sculpted away into

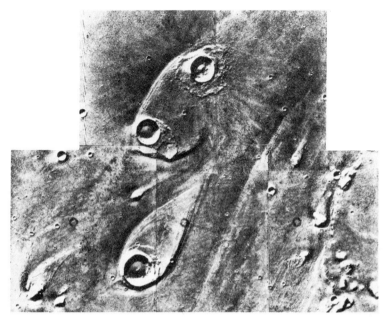

Figure 7.7.   Part of Chryse Planitia, the area selected for the Viking 1 landing site. The tear-drop shape of the 'islands' is powerful evidence for erosion by great floods of liquid water pouring down from the Martian highlands (off the picture at the bottom). Close inspection reveals that the islands are composed of many thin laminae, which have been worn away to reveal a series of terraces.

streamlined, tear-drop shapes, leaving no doubt as to which direction the seething floods came from. It would be nice to think that there were still great sheets of water on Mars, but the crater density evidence shows that the 'seas' probably dried up 2 billion or 3 billion years ago. They were probably only ephemeral anyway, the result of flash floods on a gigantic scale. If true seas had existed, one would expect to find evidence of sediment deposition rather than erosion, perhaps even basins full of sedimentary material.

Figure 7.8. This great collapse feature may have been triggered off by melting of permafrost ice at the head of the gorge; slumping of the walls has led to chaotic mounds on the floor, while, lower down, the well-defined grooves strongly suggest erosion by torrents of water released by melting ice.

The Viking cameras also gave some remarkable views of the geological process taking place within the channels and rifts. In some, huge landslides are visible beneath the walls, which are as much as 2 kilometres high in places. The landslides must have taken place on several different occasions, as some young slides can clearly be seen over-riding older ones.

The presence of fields of dunes in the bottom of some canyons shows that strong winds are at work there, and suggests that the canyons are actually being enlarged by collapse of their walls and removal of the materials by the wind – similar features can be found in desert areas of Earth. What actually causes the

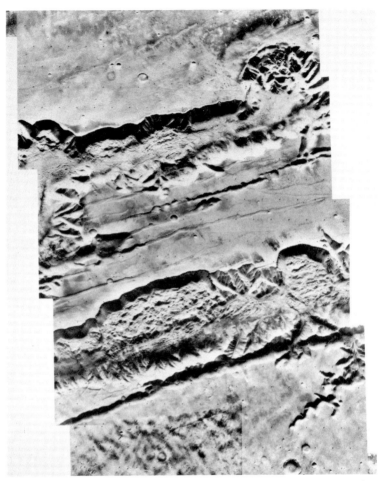

Figure 7.9. A Viking 1 close-up of part of the same giant canyon complex illustrated in Figure 6.17 (page 237). The two canyons shown are more than 60 kilometres wide, and 1 kilometre deep, and were formed initially by downfaulting along parallel faults. The troughs thus formed were then enlarged by erosion, which undercut the cliffs on the canyons, triggering off the many major slumps and slides seen. These are best displayed on the floor of the upper canyon, where several slides can be seen overriding one another.

Figure 7.10.   Several large landslides, which broke away from the cliffs at the top, can be seen in this picture. Lines of dunes are visible on the floor of the canyon, within the dark-toned area. Compare the number of impact craters on the floor of the canyon with those on the smooth highlands above.

slumping and landsliding to take place is not known; it might be groundwater sapping, or undercutting by the wind, or both. But the important point is that, while the canyons are in an area with a fairly average density of impact craters, the bottoms of the canyons have fewer, indicating that the features there must be relatively youthful.

Figure 7.11.   The first panoramic view of the surface of Mars, taken by the Viking 1 lander on 20 July 1976. The horizon is approximately 3 kilometres away, and the panorama covers nearly three quarters of the circle of view from the spacecraft. At the far left, the camera was pointed over the back of the spacecraft; towards the right, it was looking straight ahead. A better impression of the perspective may be obtained by curving the page in a semicircle and imagining yourself at the centre.

The view from the Viking landers

Although they must have been a grave disappointment for those optimists who hoped to see green bug-eyed monsters peering curiously into the camera lenses, the photographs of the Martian surface taken by the Viking landers' cameras are of superb clarity and crispness. They represent a great technical achievement. At first glance, the flat, rocky terrain in the pictures could easily be mistaken for a terrestrial desert. Angular boulders are scattered around on a gravelly surface, and small drifts or dunes of fine-grained material wind across the otherwise featureless plains which extend unbroken to the horizon. The general impression is that of a grim, sterile wasteland, but those who have travelled in deserts will recognize in the Martian landscapes the same austere beauty that they have experienced on Earth. Indeed, it is not easy

to distinguish between a picture of the Mars surface and one of the Atacama desert, so close is the similarity.

Although Vikings 1 and 2 came down thousands of kilometres apart, it is also not easy to distinguish between the two sets of pictures. Just as on Earth, a traveller in the desert often wearies of the spacious but unvarying scenery and longs for the relief of a green oasis, so any future travellers on Mars will find large parts of the planet mind-bendingly monotonous. It would not come as a surprise to any such travellers to find that the predominant hue of the surface is a dusky red. It is this uniform coloration, of course, that explains why Mars always shines down from the night sky so brilliantly blood red, and has always been identified with wars and militarism. The rich coloration is the result of chemical weathering of ferruginous material in the rocks and soil.

It might come as more of a shock to travellers on Mars to find that the *sky* is also always red, or rather salmon-pink coloured. This was something of a surprise even to the Viking science team, who had tried so far as possible to anticipate what conditions would be like. The observation itself was questioned initially, and it was suggested that the colour balance in the Viking pictures was incorrect. However, colour test charts mounted on the spacecraft gave the right results, and the correct rendering of

Figure 7.12. (opposite) *A more detailed view of the Martian surface, from Viking 2. The largest rock near the centre of the picture is about 70 centimetres across. A line of low dunes snakes across behind the rock, and partially buries several others.*

Figure 7.13. (opposite below) *One does not have to go to Mars to see thoroughly Martian scenery. This view was taken by the author in an exceedingly arid part of the Atacama desert of North Chile. The only difference is the atmospheric pressure in the Atacama is 160 times higher than that on Mars.*

Figure 7.14. (below) *The Viking lander. This is a complete spacecraft, set up in a laboratory in Pasadena, California, so that movements of the sampling arm and other components could be rehearsed before commanding the spacecraft on Mars to execute actual movements.*

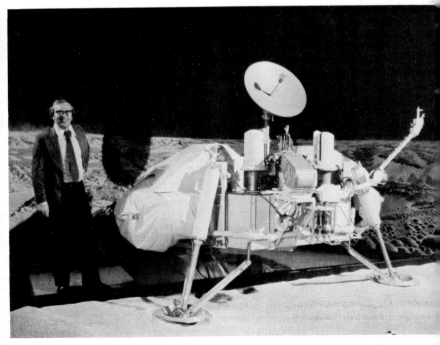

features such as orange-coloured cables left no room for doubt that, on a fine day on Mars, the Sun does shine down from a pale pink sky. The reason is not hard to find: suspended dust particles in the Martian atmosphere scatter the Sun's rays, cutting out all but the red wavelengths, just as they do on Earth at sunset. The dust particles are carried high into Mars' atmosphere during great storms, and appear to remain in suspension there for very long periods.

Apart from the all-important cameras, each Viking lander also carried a complete miniature weather station, a seismometer to detect marsquakes, a sampling arm to scoop up samples of surface soil and, most important of all, a number of analytical instruments and experiments to examine the soil, find out what it was made of, and see if life was present.

Figure 7.15.   The principal components of the Viking landers.

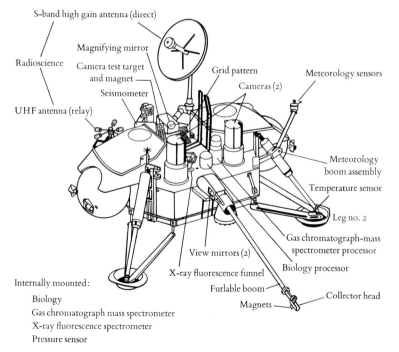

Meteorology on Mars

Although they had been designed to measure winds exceeding 300 kilometres an hour, neither of the Viking landers experienced anything other than light winds during the early part of the mission. These were nothing more than fresh breezes, blowing at less than 10 metres per second, which would cause trees to rustle and produce choppy waves on a lake on Earth; but they are nothing like strong enough to generate the lines of dunes visible in some of the pictures. There has been some controversy over these dunes, some scientists suggesting that they are ancient features, dating back to times when Mars had a thicker atmosphere, others that they are contemporary.

The possibility that dust storms might blow up while the Viking landers were on the surface presented a serious source of worry to the spacecraft designers. They feared that their precious offspring might be wrecked by the scouring, sandblasting action of the wind, or even completely buried beneath a migrating dune! At the design stage, meticulous precautions were taken to prevent wind-blown dust from interfering with the landers' mechanisms, and especially with the all-important cameras. The optical parts of the cameras were shielded, and the mechanical parts rigorously sealed. Wind-tunnel tests proved that the cameras would continue functioning until they were nearly buried. What more could one ask!

No dust storms were recorded by the cameras, but huge storms taking place in the southern hemisphere did cause the skies above the landers (both located in the northern hemisphere) to darken, and the two landers may both have been covered with a thin layer of extremely fine dust. If a storm should blow up in the northern hemisphere while the landers are still functioning, the result could be interesting, to say the least.

The ruddy red glow of the surface pictures makes Mars seem a warm, cheerful place, but the thermometers on the Viking landers revealed the bitter opposite. The minimum temperature recorded was a frigid -113 °C at the Viking 2 site, whereas the

Figure 7.16.   The Viking 1 lander's meteorology boom here rears up above the Martian landscape of boulders and dunes. No winds strong enough to form dunes were detected during the mission, and the dunes themselves appear to be suffering erosion.

maximum was a scarcely balmy − 32 °C at the Viking 1 site. As on Earth, the coldest part of Mars' day is just before dawn, and the warmest about mid afternoon.

The most memorable of all the Viking pictures were taken in September 1977, a year after the landings. The Viking 2 lander, which had touched down further north than its companion, had passed through the worst of the Martian winter when its cameras suddenly revealed a dramatic change in the scene it surveyed. During the whole of the previous year, the photographs from Viking 2 had shown the same dreary boulder-strewn landscape. Differences in shadow lengths as the Sun moved across the Martian surface provided the only visible changes. Now, however, white patches suddenly appeared covering the reddish brown soil, particularly in shadowed areas on the north side of boulders. Frost!

Figure 7.17. Although not easy to appreciate in black and white, this photograph was the most spectacular of all Viking pictures. The large boulders behind the lander cast broad shadows in which white patches of frost show up as pale-grey tones. The frost is probably a mixture of water and carbon dioxide.

This extraordinary picture did more than anything else to illustrate the complexity and fascination of Mars' meteorology. Whereas the appearance of frost was not in itself particularly surprising, since one can see from Earth how the polar caps spread equatorwards in winter, the conditions prevailing at the Viking 2 site when the frost appeared were such that it seems unlikely that the frost could consist either of straightforward water ice or of 'dry ice'. Although the temperature recorded was extremely low, calculations suggest that it would need to fall to at least $-122\ °C$ for pure carbon dioxide ice to freeze out of the atmosphere. The most likely alternative is that the frost is a compound known as a *clathrate*, consisting of six parts of water to one of carbon dioxide.

The fact that frosts are widespread in the Martian winter has one rather odd consequence. Observations of variations in the

Earth's atmospheric pressure are an essential part of the familiar ritual of weather forecasting. If, on tapping his barometer one morning, a meteorologist were to note a drop in pressure from 1,000 millibars to 998·5 millibars overnight, he would think nothing of it. But if this drop were to be repeated every night for weeks until the barometer stood at only 700 millibars, he would (apart from having difficulty with his breathing) be confused and perplexed, because it would mean that the Earth had somehow lost 30 per cent of the mass of its atmosphere. Such huge variations are a regular part of Mars' meteorology. Over a period of about 100 Mars days, both Viking landers recorded a slow and steady decrease in atmospheric pressure. At the Viking 1 site, the pressure dropped from 7·6 to 6·8 millibars, a difference of about 10 per cent. Over the course of a full Mars year, a variation of nearly 30 per cent is forecast. The reason is straightforward. As winter advances in one hemisphere, carbon dioxide (and water) are progressively frozen out of the atmosphere, reducing its mass by 30 per cent at maximum. When the weather warms up again, the carbon dioxide sublimes, and the pressure rises again. Of course, events in one hemisphere are the reverse of those in the other, but the net effect is that the planet's atmospheric pressure is directly affected by the total amount of ice on its surface.

The atmosphere

Apart from making meteorological observations, the two Viking lander craft were also capable of analysing the atmosphere, both at high altitudes during their descent and at the surface. These measurements confirmed that the atmosphere is composed predominantly of carbon dioxide, with small amounts of nitrogen, argon, oxygen and carbon monoxide, much as had been expected. The bulk composition of Mars' atmosphere, then, is broadly similar to that of Venus, but in its trace element and isotopic composition it is unique.

Two factors stand out. First, Mars is relatively rich in the isotope of argon called argon 40, and relatively poor in the isotope argon 36. There seems, in fact, to be a progressive decrease in argon 36 with distance from the Sun. This is not easy to understand, since argon 36 is more volatile than argon 40 and there are reasons (discussed in the final chapter) for supposing that there should be an *increase* in volatile components away from the Sun.

Secondly, Mars' atmosphere is rich in the isotope nitrogen 15 – in fact, nitrogen 15 is 75 per cent enriched relative to nitrogen 14 in the Earth's atmosphere. This has been interpreted as meaning that the lighter isotope (nitrogen 14) has preferentially escaped from Mars' atmosphere with the passage of geological time. Since the initial nitrogen 15 to nitrogen 14 ratio was presumably originally close to that of the Earth, this means in turn that there must at one time have been far more nitrogen in Mars' atmosphere than there is now. Estimates vary, but some suggest that it was ten times more abundant than it is now, and may itself have exerted an atmospheric pressure equivalent to about 8 millibars.

This is interesting enough in itself, but there is more, much more. Two American students of planetary atmospheres, Edward Anders and Tobias Owen, have used the data on nitrogen and argon isotopes in Mars' atmosphere to work out 'release factors' for other volatile gases, and thus make estimates of water and carbon dioxide that might once have existed on Mars. They reach the startling conclusion that the pressure accounted for by carbon dioxide may once have been as much as 140 millibars, and, even more amazing, that enough water vapour could have existed to provide a layer of liquid 9 metres deep over the entire planet. As they laconically observed in their paper, with such a volume of water available 'we should have the wherewithal for cutting the famous dendritic channels'. A further, less obvious implication of large amounts of water vapour and carbon dioxide is that the Martian climate would have been a lot warmer, since the two gases together would form an excellent greenhouse. Given plenty of water, and mild temperatures, life *might* have evolved . . .

Anders' and Owen's conclusions are not universally accepted, and there is bound to be much future debate about how Mars' atmosphere has evolved. Their work does at least demonstrate, however, the fundamental implications that can be drawn from a few dry, apparently uninspiring numbers.

The rocks and soil

To a geologist, the Viking surface pictures are tantalizingly clear and sharp. When the rocks appear so invitingly, it is frustrating not to be able to pick one up and hammer it! And although the Viking landers were equipped with sampling arms capable of scooping up handfuls of soil for analysis, this is *not* the same as giving a boulder a hefty swipe to see what is inside it. It is also difficult to be certain about exactly what one is seeing in the pictures. For example, almost all of the boulders seem to be pitted and hollowed on the outer surfaces, giving them a spongy appearance. The holes could be original gas bubbles (vesicles) in volcanic lavas, or they might be the result of weathering of the rocks, where the wind has preferentially eroded away some minerals and not others.

Figure 7.18.   A high-resolution photograph of spongy looking rocks near the Viking 2 lander.

Again, while several different types of rock *appear* to be present, it is impossible to know what the differences are due to – they may be merely superficial. Geologists know only too well the bewildering range of disguises an ordinary rock like limestone can adopt when subjected to different conditions of chemical and physical erosion, and simple surface staining. It is not even possible to determine where the rocks at the Viking landing sites came from. At one extreme, they might be fragments ejected by successive meteorite impacts on the smooth plains, while at the other they might have been eroded out of an older deposit by some great flood early on in Mars' history and transported for long distances before coming to rest.

The evidence from the Viking 1 site favours the first of these alternatives. The photographs show flat outcrops of bare, solid rock – so the boulders which dominate the landscape could have been derived locally. Furthermore, the unmistakable profile of a crater rim can be seen on the horizon only 2·5 kilometres distant, and three others are known to be located equally close to the

Figure 7.19. This spongy looking rock in the Atacama desert is of volcanic origin, and its spongy appearance is caused solely by wind erosion, which has preferentially attacked relatively soft feldspar crystals.

lander. At the Viking 2 site, no craters are visible, and there is no solid rock exposed. It is possible, however, that the site lies within the ejecta blanket of a 100-kilometre diameter crater called Mie, situated 160 kilometres to the east.

Apart from the boulders, the most eye-catching features of the lander pictures are the dunes. Close examination shows that these are by no means simple sand dunes, such as we are familiar with on terrestrial coasts and deserts. Sand dunes consist of quartz grains about half a millimetre in diameter. Although this is not large, the particles are quite big enough to make their presence felt when the wind whips them against bare legs and ankles, as anyone who has walked on beach dunes in windy weather can testify.

On Mars, the dunes do not consist of quartz grains, and the grains themselves are much smaller, comparable with terrestrial silt and clay grain sizes. Furthermore, while one can actually perceive the steep front face of a sea-shore dune steadily slipping and sliding forward as the wind carries fresh grains over its crest, the Martian dunes appear not only to be static, but also to be undergoing active erosion. Many of the dunes so sharply portrayed in the Viking 1 pictures show unmistakable traces of internal stratification on surfaces which are themselves sculpted by the wind. Such internal stratification is *only* seen in eroded dunes; active dunes show only their uniform exterior.

Thus, the Martian dunes in the photographs are unquestionably at present in the process of retreat. It is impossible to deduce from the pictures how extensive they were formerly, or when they were constructed. It may be that they are rather ancient relics of a much thicker sediment deposit. At this stage in our untangling of Mars' geological history, we simply do not know.

The first analysis of a soil sample at the Viking site was the first of any planetary material, not counting meteorites recovered on the Earth and the samples from the Moon. The results showed that the soil consists mostly of a mixture of silicate minerals. This discovery was not unexpected – it means that the minerals making up the soil are basically the same as those found in terrestrial soils; there is nothing bizarre or utterly outrageous.

The analysis also confirmed that the soil is rich in iron, probably in the form of ferric oxide which coats the grains and gives the characteristic rusty-red appearance to the Martian landscape. This rusty pigment, however, appears to be only a thin film, perhaps less than 2 microns thick.

A more surprising discovery was that the soil is rich in sulphates and carbonates. It is highly unlikely that the sulphur is present in its elemental form; it is much more probable that it is present in the form of hydrated sulphate minerals, since about 1 per cent of water is present in the soil. Sulphate minerals are highly soluble in water, and thus there is no likelihood of finding them in the soil of the average terrestrial garden. However, they *are* quite common on Earth, and consideration of the environments in which they occur here gives a guide to conditions on Mars.

The commonest terrestrial sulphate mineral is *gypsum*, calcium sulphate. This is the mineral that forms the bulk of all the so-called 'salt' flats in the world: the glistening white flat surfaces characteristic of basins of internal drainage in desert areas. The most important feature of these salt flats is not burning heat, as is commonly supposed, but intense aridity. The biggest salt flats on Earth are high up in the chill air of the Andes, where the night-time temperatures are regularly tens of degrees below zero; but, during daylight hours, solar radiation is so intense and the air so dry that the rate of evaporation is colossal. This means that there is a steady upward movement of ground water to the surface. As it moves upwards, it dissolves or leaches out all the more soluble components of the rocks and soil; but when it reaches the surface, it evaporates and deposits its burden of dissolved salts. Thus, in the Andes and other arid regions, the surface of the desert often consists of a distinct hard crust known as a *duricrust*, consisting of leached soil and soluble salts deposited by evaporation – the great Chilean nitrate fields were formed in this way.

A similar process seems to have operated on Mars, though it is less certain from where the water originated. It may have been released by melting of permafrost ice in small quantities over long periods. An important consequence of all the leaching and weathering that has taken place in the Martian soil is that it is difficult to work out the original chemistry of solid rock particles. Comparisons with soils formed by the breakdown of different terrestrial rocks suggest that the Martian source rocks were probably basic igneous rocks, similar to basalts – a conclusion which fits well with other observations from the Viking orbiters and landers. Some of the lander pictures show boulders honeycombed with bubble-like cavities. These could well be vesicles produced by gas escaping from a molten basalt lava flow.

The search for life

The ultimate goal of the Viking missions was to try to answer the burning question: 'Is there life on Mars?' The scientific data obtained prior to the missions had shown to the satisfaction of all but a few diehards that the chances of finding organisms of the size and complexity of those on Earth were remote, but that much simpler forms just *might* exist. There were four main factors mitigating against large Earth-like animals being at large. First, the atmosphere is tenuous, and deficient in oxygen. Secondly, liquid water cannot now exist on the surface. Thirdly, temperatures are so low that metabolic rates must be very slow. Fourthly, the surface is exposed to blistering ultraviolet radiation from the Sun.

Those who wanted there to be life on Mars could easily put up arguments to counter these apparently insuperable problems. Many of these are perfectly rational. The lack of oxygen is perhaps the easiest to deal with. Many simple terrestrial organisms (bacteria) thrive without oxygen, and, indeed, are poisoned by it. Some of these bacteria live in sulphate-rich environments, chewing up the sulphates, reducing them to sulphides, and giving off oxygen.

The lack of water is more difficult to counter. Every terrestrial organism requires water for its metabolism. It was argued, however, that organisms might exist in the Martian soil which are capable of using water locked up in permafrost, or, more plausibly, that they might flourish in isolated 'oases' where volcanic heat was sufficient to keep supplies of liquid water available. Many separate 'oases' of life might exist on Mars; the chances of the Vikings landing near them were slim.

A corollary of the lack of water is the low temperature. It was at first thought that the best places to land the Vikings would be near the equator, where it would be warmest. However, it was soon realized that these areas are also the *driest*, so eventually a compromise was reached, and Viking 2 was landed at a relatively high latitude where the water-vapour level was highest.

Finally, it was well known that on Earth human beings living at high altitudes are exposed to much higher levels of ultra-violet radiation than those living at sea level; they also show a much greater incidence of skin cancers. The Earth's atmosphere acts as a thick, protective blanket, cocooning us with gases (especially oxygen and ozone) that interact with the ultraviolet rays and prevent them reaching the surface. Any Earth-like organism on the surface of Mars would be exposed to the full glare of the Sun, and would be literally sunburned to death.

It was, of course, possible to rationalize away this and the other problems in one fell swoop, and to breed (in the imagination) organisms that would positively thrive on Mars. Such organisms would live in Mars' soil in areas of volcanic heat, where plenty of liquid water would be available; would not breathe oxygen but would metabolize carbon dioxide like plants; might enjoy supping on sulphates, and would protect themselves from ultraviolet radiation by evolving thick carapaces of opaque, stony material. Protected by such a stony parasol, a Martian organism would be able to pop its head up above the surface from time to time to take in the view, or even wander around on the surface.

Although they realized that such hypothetical organisms were more the offspring of fevered imaginations than sound science, and that the dice were loaded against even the humblest life forms existing on Mars, the Viking biologists left nothing to chance, and devised experiments capable of detecting *any* kind of organism, from elephants to bacteria.

As soon as the first pictures from the landers were received, it was obvious that there were no bizarre bugs bicycling around, or holding up placards saying 'Go Home Yanks', as some cartoonists had facetiously suggested. More seriously, when used in a specially designed fixed scan mode, the cameras detected no signs of *any* movement whatever, eliminating the possibility of there being even tiny organisms at large at the landing sites, much less fancy creatures wearing stoney sunhats. However, as the Viking scientists themselves commented with true scientific caution, their cameras scanned only two sites, covering less than 0·0000004 per cent of the surface of Mars, so that the possibility of large mobile life forms cannot be ruled out – it would be like saying that, because there are no elephants in the Sahara or in the middle of the Antarctic ice-cap, there is no life on Earth.

Although the cameras themselves could not finally prove or disprove the existence of life on Mars, the Viking biologists hoped that the life-seeking experiments on board the spacecraft would be able to provide unequivocal answers. As it turned out, the data that came back were far more ambiguous than the experiment designers had thought possible. A lot of heads were scratched when the data first arrived, and these heads were busy for months afterwards, trying to work out what the data meant.

The simplest result came from the instruments called the gas chromatograph mass spectrometers (GCMS). These instruments looked for organic compounds in the soil samples; they were capable of detecting organic molecules at a level of only a few parts per *billion*, and similar ones had quite happily detected twenty or more different organic compounds in soils from the Antarctic, the most sterile terrestrial soils known. Even these contain a few hundred micro-organisms per gram.

On both the Viking 1 and 2 landers, the GCMS instruments found nothing. Absolutely nothing.

This is a most serious obstacle to those who want there to be life on Mars. Unlike most of the other experimental data, it is not easy to rationalize it away. Only one explanation is even worthy of consideration, and this is that, whereas the GCMS detects organic material contained in both living organisms and the dead material left by earlier ones, it is possible that the organisms on Mars are such efficient scavengers that they vacuum up *all* the organic debris of their predecessors, and their metabolic products. Furthermore, the actual mass of living material at any one instant must be quite tiny.

There are many serious objections to this tempting picture of cannibalistic Martian organisms. For one thing, it was confidently expected that the GCMS would detect organic compounds of non-biological origin. There are several possible sources of these, but the most important is probably in outer space. Some kinds of meteorite contain carbon and organic compounds, and these must certainly be falling on Mars. The fact that organic compounds from this source were *not* found demonstrates that some vigorous process is at work to destroy them. It is much more probable that this is an inorganic chemical or physical process than a biological one – strong ultraviolet radiation by itself could be sufficient.

The other three biology experiments came up with even more puzzling results, and one of them strongly indicated some kind of biological activity, but did not prove it unequivocally. All three experiments were conceived along the same lines: take a soil sample with the presumed Martian micro-organisms in it, put the soil in a favourable environment, supply the bugs with good things to eat and drink, and watch what happens.

In the *pyrolitic release experiment*, small samples of Mars soil were exposed to a Mars-like atmosphere containing carbon monoxide and carbon dioxide gases which had been 'labelled' with the radioactive isotope of carbon, carbon 14. The samples were then kept warm for five days, and illuminated by Mars-like sunlight, but without the harmful ultraviolet rays. The gases were then purged from the chamber, and the soil heated strongly. The idea was that if any Mars bugs were present in the soil, they would become active when exposed to the warm and cheery environment of the test chamber and use up some of the labelled gases, converting them by photosynthetic processes into organic compounds. The strong heating would then have destroyed these compounds, and the labelled gases would be released once more into the test chamber, where the radioactivity from the carbon 14 could be easily detected. Since the unused gases had been purged,

Pyrolytic release experiment

Labelled release experiment

Gas exchange experiment

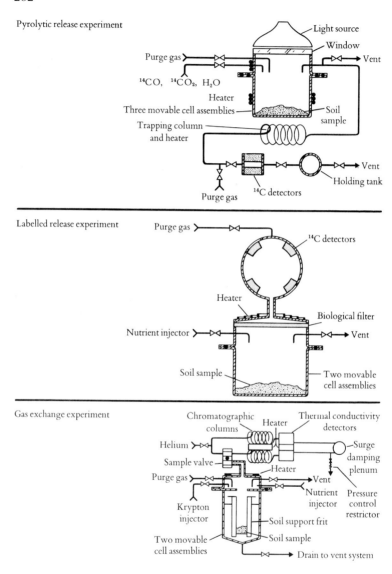

Figure 7.20. A diagrammatic summary of the Viking biology experiments.

any labelled gas detected *after* the strong heating would presumably have been used by an organism in its metabolic processes.

So much for the theory. In practice, the experiment showed that a small quantity of labelled gas was indeed converted to organic material, and, most importantly, when this process was repeated with a 'control' sample that had been pre-heated to 175 °C to sterilize it, no such gas conversion took place. At first sight, this seemed strong evidence for biological activity, but when the experiment was repeated with a sample which had been heated to 90 °C, hot enough to kill almost all terrestrial organisms, the *same* result was obtained as with the unheated sample. Thus, since GCMS had failed to detect *any* organic compounds to start with, it seemed more likely that some kind of non-biological process must have taken place to account for the breakdown of carbon monoxide and dioxide.

In the *gas exchange experiment*, much the same procedure was carried out, except that here the soil samples were placed in a test chamber under optimal conditions and supplied with a rich nutrient brew. The gases above the sample were monitored continuously by a process known as gas chromatography to detect changes taking place in them owing to organisms in the soil consuming some of the brew, and releasing gases as a result of their metabolic processes. The experiment was designed to detect small numbers of slow-growing micro-organisms, but it gave *no* indication of positive results on Mars. Even after incubation periods of *seven months*, the gases in the test chambers showed no changes attributable to biological processes.

The third experiment, which gave some highly controversial results, was known as the *labelled release experiment*. The soil samples were again placed in a test chamber under optimal conditions, and again supplied with a rich liquid nutrient brew, in which the carbon compounds were labelled with radioactive carbon 14. If bugs existed in the soil, it was hoped that they would accept the nutrient brew as eagerly as most of us would accept a free lunch. Subsequently, as their metabolic processes got to work, they would liberate labelled carbon dioxide or methane gases into the test chamber, and these would then be easily detected. It is just like giving someone a radioactive free lunch, and then asking them to belch into a geiger counter.

In the actual experiments, fresh samples of Martian soil were incubated in the nutrient brew for periods up to fifty-one days. When the soil and brew first came into contact, there was an immediate release of substantial amounts of labelled gas. The rate of release subsided rapidly, but over subsequent weeks, labelled gas continued to be released slowly. Pre-heating of the soil to 160 °C in a control sample completely inhibited the release of labelled gas, while heating to 50 °C cut down the rate of release by about 65 per cent.

Taken in isolation, these observations are strongly suggestive of biological activity. But, of course, they cannot be taken in isolation. If anything, the negative results from the GCMS are an extremely strong pointer to the absence of life on Mars, since, even if there were no organisms at all *living* at either of the landing sites, it is scarcely credible that the instruments would not have detected minute quantities of biological materials blown by the wind from other parts of the planet.

While it is possible to explain *all* the results from the bio-chemistry experiments by devising rather odd kinds of Martian organisms to fit them, it seems much more likely that the surface inorganic chemistry of Mars is distinctly odd, much odder than the Viking scientists had at first thought. Many laboratory experiments were carried out on Earth in the wake of the Viking landings to see if the landers' results could be duplicated here. One strong possibility to emerge from these studies is that Mars' soil includes vigorously oxidizing compounds, such as peroxides and superoxides, which themselves react vigorously (and quite non-biologically) when brought into contact with aqueous liquids. Such compounds might result from the interaction of the intense ultraviolet radiation from the Sun with the atmosphere and soil, and would create a chemically hostile environment lethal to any *known* form of life.

Even if this could be proved to be the case, there would still be a few diehards remaining who would say that it was a waste of time anyway to look for *known* forms of life on Mars. Martian organisms, they would argue, are so impossibly bizarre that no instruments designed by mere Earthlings could ever detect them. Perhaps their biochemistry is based on silicon, not carbon? Some people take a lot of convincing!

The deep interior of Mars

The extraordinary success of the Viking missions was marred by only one failure. Although the two landers were designed primarily to search for life, the potential scientific value of information on the internal structure of Mars was so great that each spacecraft was equipped with a seismometer to detect any 'marsquakes' occurring during the period of the missions. The instruments were not as sophisticated as those carried on the Apollo missions, and suffered the additional disadvantage of being carried in the body of the spacecraft rather than being placed firmly on the ground. Thus, there were numerous sources of interference, ranging from the vibrations caused by movements of the sampling arm to gentle shaking by the wind. Notwithstanding these problems, the seismometers were capable of providing the first direct evidence of the internal structure of another planet.

Sadly, things did not quite live up to expectations. The sensitive instruments had to be 'caged' during their long flight from the Earth to prevent them from being damaged by the jolts and jerks of launch, orbital decoupling and landing. For some reason, the Viking 1 seismometer failed to uncage when commanded to do so from Earth. This instrumental failure was enormously frustrating to the waiting seismologists on Earth, who must have felt rather like Jerome K. Jerome's characters in *Three Men in a Boat* who, having rowed hard for hours in eager anticipation of a picnic lunch, were faced with a tin can of most succulent pineapple but did not have a tin opener to get at it. The failure of the Viking 1 seismometer was all the more important, because it devalued the results of the Viking 2 instrument, since without it there was no means of cross-checking data.

The Viking 2 seismometer worked perfectly. Over a period of

171 Mars days, it patiently recorded every tiny tremor that shook the spacecraft. The vast majority of these were generated by the wind; the remainder were caused by movements within the spacecraft itself, such as those of the cameras. Only one 'event' was identified which could have been a 'marsquake', but it might also have been caused by a meteorite impact. The lack of seismic events recorded by Viking 2 is an important scientific observation in itself; it indicates that Mars is a seismically 'quiet' planet. It may well prove to be as inactive as the Moon, but much more data will be required to determine this. Since a new generation of spacecraft landing at several different sites will be needed, it will be many years before it is known for certain how seismically active a planet Mars is.

Lacking any seismic data to probe the interior, our knowledge of Mars' deep structure has to remain hypothetical. Fortunately, quite firm constraints can be placed on the possible internal structure from knowledge of the planet's density, rotation period and the extent of its flattening – Mars is significantly more 'squashed' in shape than the Earth. Current thinking suggests that Mars has a three-fold structure of core, mantle and crust like the Earth's, but that each of the three components is different from the Earth's.

Because Mars is much less dense than the Earth, its core cannot contain so much metal. The core is probably made up mostly of iron and iron sulphide, and has a radius of about 1,700 kilometres. There are some reasons for thinking that the core might be molten, but since Mars lacks a significant magnetic field, this is uncertain. The mantle of Mars may, strangely enough, be *denser* than the Earth's, but this is balanced by the fact that it forms a smaller proportion of the total volume of the planet. The higher density is due to the fact that the Martian mantle contains more iron oxide than does the Earth's.

The composition of Mars' mantle may seem to be of little immediate interest, as dull and desiccated as the problems that philosophers pose themselves concerning the numbers of angels that can dance on a pin-head. Mars' mantle, however, is the key to the geology of the planet: once we understand the mantle, we can understand the whole planet. For example, it has been suggested that Mars' denser mantle may mean that it gives rise to the eruption of far more fluid basalt lavas than are spewed out by volcanoes on Earth. If this is the case, then it becomes easier to interpret the vast areas of 'smooth plains' covered by lava flows, and the maze of thin flows that surround volcanoes such as Olympus Mons. Furthermore, it is critical to know whether or

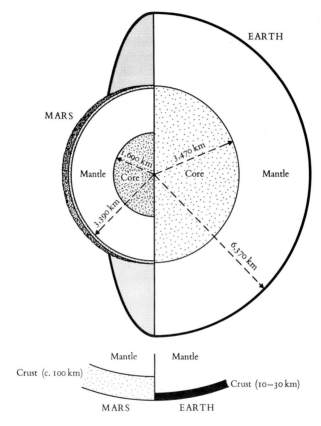

Figure 7.21. The internal structure of the Earth and Mars compared. The Earth's core is about the same size as the whole of Mars (above), but the crust of Mars is much thicker than the Earth's (below).

not there is a molten layer within the mantle: if there is, then there should be at least some evidence of active tectonic processes at the surface.

Because no samples have yet been returned to Earth from Mars for age dating, it is not yet possible to outline a firm history of Mars. A broad outline of its evolution, however, has been proposed by M. N. Toksoz and A. T. Hsui of the Massachusetts Institute of Technology, by considering what might have happened to Mars' internal heat.

At about 4·6 billion years ago, the planet formed from the solar nebula. Between 4·6 and 4 billion years ago, the planet heated up, as a result of the accretion process and through decay of short-lived radioactive elements. The core separated out in a molten form, and the crust began to form at the surface. This process continued until about 3 billion years ago, by which time a thick crust had formed. Between 3 and 2 billion years ago, the internal heating of the planet steadily increased, and expansion of the interior caused tensile features to open up at the surface of the planet. In the subsequent billion-year period, geological activity on Mars probably reached its peak, with vast volcanic eruptions forming the huge plains, the great rift valley, Vallis Marineris, gaping open, and large volumes of gas being blown out from the mantle into the atmosphere. This, in turn, may have led to the production of a dense atmosphere, which may have coincided with one of the periods in which the great channels were carved.

About 1 billion years ago, the mantle began to cool. The crust got steadily thicker, reaching its present 200 kilometres and thus making it more difficult for volcanic magmas to break through to the surface. Volcanic activity therefore tapered off. In the last few hundred million years, the cooling has continued. Probably, the molten part of the mantle is now more than 300 kilometres beneath the surface. Thus, although volcanic activity may still continue at a feeble level, it is all but finished on Mars. Similarly, it is clear that tectonic processes never got as far on Mars as they did on Earth, and that they are now winding down.

Mars, it seems, is not a dead planet, but a dying one. The science-fiction writers were not so far from the truth, after all.

Phobos and Deimos: a final word

Even in the most powerful telescopes, Mars' two tiny moons appear only as specks of light. Mariner 9 succeeded in photographing both Phobos and Deimos and thus made it possible for us to look for the first time upon the surface of satellites other than our own. The Viking missions provided much better pictures, as the result of some exquisitely calculated trajectory changes. At one point, the Viking 2 orbiter swept past Deimos at a distance of only 26 kilometres. The navigational control was so precise that it would have been perfectly possible to land the spacecraft, had it been equipped to do so.

The close-up pictures revealed that the two moons are irregular lumps of rock, tumbling through space like pock-marked potatoes. Phobos has an ellipsoidal shape, with a greatest dimension of 27 kilometres, while Deimos is smaller and rounder, with

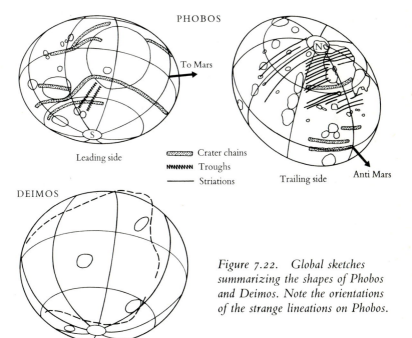

Figure 7.22. Global sketches summarizing the shapes of Phobos and Deimos. Note the orientations of the strange lineations on Phobos.

a greatest dimension of 15 kilometres. Like our Moon, both Phobos and Deimos exhibit spin-orbit coupling; they are tidally locked, so that in each case their long axis always points towards Mars. Phobos whips round Mars in 7 hours 39 minutes, Deimos, further out, wanders round in a more leisurely 30 hours 18 minutes.

The most surprising feature of these miniature moons is that they are heavily cratered. Phobos in particular has one crater, named Stickney, which is so large that the impact which produced it must have come within an ace of knocking the little moon to pieces. Phobos also has an extraordinary lineated appearance, so that it looks at first sight as though it has been wound round with wool. Nothing like these lineations has been seen elsewhere in the solar system. They appear to consist of parallel grooves in the

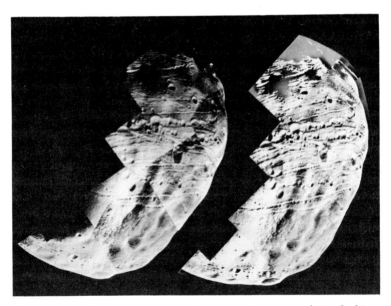

Figure 7.23.   One of the most amazing space pictures ever obtained, this view of Phobos was obtained by the Viking 1 orbiter when it flew past at a distance of less than 300 kilometres. The linear grooves are unmistakable. The picture on the right is a computer-enhanced version of the one on the left.

surface, along which are located lines of craters. The grooves are typically 100–200 metres wide, and 20–30 metres deep.

In some way, the grooves appear to have been associated with the formation of Stickney, since they are best developed near the crater. One suggestion is that the impact opened up linear cracks into which loose rocky material then fell; another is that loose material was ejected by gases escaping through the cracks and excavating the craters which are aligned along the surface.

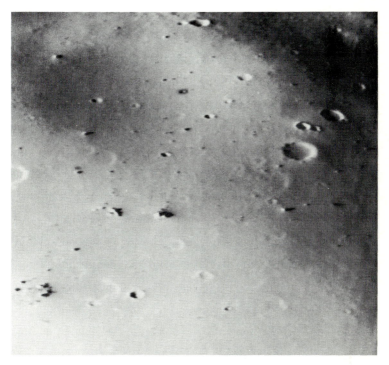

Figure 7.24.   Not so eye-catching as Figure 7.23, this picture of the surface of Deimos is just as impressive in its own way. It was taken from a distance of only 50 kilometres, and shows objects as small as 3 metres across: an automobile parked on the surface of Deimos would show up clearly. The large boulders are 10 metres across, and were probably ejected from near-by craters.

It is possible that Phobos' unique grooves are a consequence of the fact that both moons are made of weak, unconsolidated material, not solid rock. This is indicated by their low density (about 3×10^3 kg m^{-3}), and by their extremely low albedo – both are very much darker than our Moon. These observations strongly suggest that the two moons are made of carbonaceous material, similar to that found in some rare types of meteorite.

It seems certain that Phobos and Deimos became moons of Mars early in its history. They may be scraps of material left over from the formation of the planet, or they may be odd lumps picked up in its orbit through a solar system that was a good deal more cluttered with rocky fragments than it is now. The key to their origin probably lies in their carbonaceous composition, as will become apparent in the final chapter.

Figure 7.25.   The silhouette of Phobos seen against the surface of Mars. Phobos is one of the darkest objects in the solar system; hence, in this picture, correctly exposed for the Martian surface, it appears black. Its odd shape is obvious.

8. Jupiter: the Giant Planet

Jupiter is not so much a single planet as a small solar system in its own right. Not only is its mass greater than the rest of the planets put together, but it also has at least fourteen satellites dancing attendance on it, two of which are as big as the planet Mercury. Jupiter is therefore qualitatively different from the inner planets that have so far concerned us. The inner planets were much easier to deal with. Because they are made mostly of rocks, they are fairly Earth-like and are capable of description in Earth-like terms. Spacecraft pictures of Mars show mountains, craters and valleys which are entirely alien in some respects but quite familiar in others.

Jupiter, and the other outer planets, are remote from us in every way. Apart from their great distances from the Sun, they do not even have solid surfaces and consist largely of liquid gases. Some of the phenomena that can be observed in the extreme outer parts of their atmospheres can be related to terrestrial atmospheric processes, but we can only guess at what goes on at deeper levels. Although the major planets therefore present considerable difficulties in interpretation, their many satellites are much more accessible to simple observation, and have already yielded some of the most startling discoveries of a decade that has been rich in revelations. The satellites have been, and will remain, just as important targets for spacecraft exploration as their parent planets because they are easier to investigate, and also because, in their sheer number and variety, they are bound to contain important clues to the evolution of the solar system.

Jupiter orbits the Sun at a mean distance of 778 million kilometres, three times further out than Mars. Few spacecraft have yet completed the long voyage from Earth. Pioneer 10 successfully traversed the asteroid belt and flew past Jupiter in December 1973, and Pioneer 11 followed a year later. Both obtained superb pictures of the planet and its atmospheric complexities, but only transmitted back tantalizingly fuzzy images of the satellites. Pioneer 11 continued its voyage past Jupiter towards Saturn, and arrived in September 1979.

The Pioneer missions are unique in one respect. The two spacecraft will be the first man-made objects to leave the solar system altogether. They were travelling through space while these words were being written; they will still be drifting on when they are read. If, by chance, anyone should stumble across a copy of this book in 100 or 1,000 years from now, the Pioneers will still be travelling onwards, plunging further and further into the oblivion of deep space. At some unimaginably remote point in the future, there is a tiny possibility that one of the spacecraft will drift into a part of the galaxy in which other intelligences live. They may – if they are interested – investigate the spacecraft and try to determine where it came from. They will find a small plaque thoughtfully provided by the NASA scientists, containing information on the location of Earth, and pictures of a man and a woman. An appealing gesture, perhaps, but a futile one, because Man will certainly be extinct before his message is read.

Following the Pioneers, no new missions to Jupiter were launched until 1977. In that year a pair of Voyager spacecraft were launched which were planned to fly past both Jupiter and Saturn, obtaining higher resolution pictures of the planets and their satellites than the Pioneers, and making a number of other observations, particularly of the planets' magnetic fields. Voyager 1 successfully flew past Jupiter in March 1979 and was expected to reach Saturn in November 1980. Voyager 2 flew past Jupiter in July 1979 and should reach Saturn in August 1981. If all goes well, this epic voyage of discovery will culminate in 1986 with the first fly-by of Uranus.

Size, shape and density

Jupiter is such a big planet that it does not take much in the way of a telescope to show its yellowish, striped disc. Its basic statistics are therefore relatively easy to determine. Its equatorial diameter is a whopping 142,800 kilometres (11 times that of the Earth), but its polar diameter is only 134,000 kilometres. Jupiter therefore definitely resembles an orange that has been sat on, and this 'squashing' is clearly visible through a telescope and in pictures.

The Earth and Mars also show tiny amounts of squashing. It is a consequence of the forces produced by their rotation. But why is the effect so marked with Jupiter? This is easily answered by measurement of Jupiter's rotation period. Careful clocking of visible features shows that they come round every *9 hours 50 minutes*, a remarkably short period for such a large planet. Clearly the centrifugal forces which create the equatorial bulge are far larger in a planet which is both much bigger than, and rotating twice as fast as, the Earth.

Measurements of the rotation period of visible features on the surface of Jupiter reveal another peculiar fact: the apparent rotation period varies from place to place, and is generally longer nearer the poles than at the equator. This means, of course, that the features observed are merely atmospheric manifestations; they are not 'attached' to any solid part of the planet. This gives a pointer to the powerful atmospheric disturbances which exist on the planet, of which more later.

The *mass* of Jupiter is also easy to work out, since it has many satellites. It is about 1.9×10^{27} kg. That sounds a lot, and indeed it is – the *total* mass of all the other planets put together is only about 7.6×10^{26} kg.

The *density*, by contrast, is low, at about 1.3×10^3 kg m^{-3}, indicating that whatever Jupiter is made of, it must be quite different from the material making up the inner planets. If Jupiter *were* made of the same materials, it would be quite a planet!

The internal structure of Jupiter

It is not easy to determine what Jupiter is made of. Merely observing it does not help much; the magnificent pictures taken by Pioneer and Voyager spacecraft show only the tops of towering clouds in the enormously deep atmosphere. There are, however, a number of important guidelines. First, there is the density. Jupiter *must* be made up predominantly of light elements, and *cannot* contain much in the way of metals or silicate minerals that form rocks – these are simply too dense. Since hydrogen and helium are the most abundant elements in the solar system, and both are light gases, it is reasonable to assume that these are present in large quantities.

Next, there is Jupiter's distinctly flattened shape. Now, while a degree of oblateness is to be expected in a rapidly rotating planet, oblateness is also controlled by the distribution of mass within the planet. If most of the mass were concentrated in the outer parts of the planet, it would have a much more ellipsoidal shape than if most of the mass were at the centre. Theoretical calculations for Jupiter show that its flattening is much less than would be expected for a homogeneous planet, and thus it must have a concentration of mass towards its centre. Similar calculations also show that the pressure at the base of Jupiter's 1,000-kilometre thick atmosphere is great enough to liquefy hydrogen; at depths of about 25,000 kilometres, it is great enough to convert 'ordinary' liquid hydrogen into a metallic form.

Metallic hydrogen sounds hard to swallow. (Indeed, it would be!) But it is not really an outrageous suggestion, since all gases can be converted to solid forms by applying sufficient pressure. In the case of hydrogen, this was first done in 1978 at the Carnegie

Institute of Washington, when a small quantity of intensely cold liquid hydrogen was trapped in a small cavity between two diamonds in a special high pressure cell. The cell was allowed to warm up to room temperature, which caused the trapped hydrogen to try to expand, and thus to exert enormous pressure on itself. A microscope focused through the diamond enabled observers to see the hydrogen solidify at a temperature of 25 °C, and a pressure of about 57,000 atmospheres – equivalent to that of a column of mercury 43 kilometres high!

Further warming caused the pressure to increase to about 360,000 atmospheres, comparable with that deep inside Jupiter. The hydrogen remained solid, but appeared to switch over to a different, higher-pressure type whose refractive index increased sharply with increasing pressure. The increasing refractive index implies an increase in density of the solid hydrogen, which, at a pressure of 360,000 atmospheres, was about 0.7×10^3 kg m^{-3}. In this state, solid hydrogen has many of the properties of a metal.

Now it is a vast step from solid hydrogen in a tiny diamond pressure cell to the centre of Jupiter, but the experimental results help to demonstrate that massively compressed hydrogen is likely to form a large part of the low-density 'giant' planets. Much still remains to be learned and understood, however, especially since the centre of Jupiter is unlikely to be *solid* hydrogen, because the temperature there is far too high, perhaps as much as 11,000 °C. In these conditions, it is thought that hydrogen may exist in a strange form that is both liquid *and* metallic; that is, electrically conducting. In this form, the hydrogen probably exists as dissociated atoms rather than molecules.

It is essential that the material making up the bulk of Jupiter's mass be electrically conducting in order to explain another important property of the planet, its magnetic field. Jupiter has an extremely strong magnetic field, ten times more powerful than the Earth's. The field is broadly dipolar (which means it has north and south poles, like a bar magnet), but it is much more complicated near the surface, where quadrupolar and octopolar fields were detected by Pioneer 11. Incidentally, Jupiter's north magnetic pole is at the *bottom* of the planet (the south *geographic* pole). This would be a bit confusing for anyone trying to use a compass on Jupiter, but it is worth recalling that the Earth's north magnetic pole was also at the bottom of the planet about a million years ago – the field reverses itself every now and then.

The significance of Jupiter's magnetic field for its internal structure is this: to generate a magnetic field in a planet requires, first, an electrically conducting material, and, secondly, a source of energy. In the Earth, the electrically conducting material is the liquid iron in the core, and the magnetic field is generated by thermally driven currents in the molten iron. These interact to form a kind of dynamo.

Jupiter's density does not allow for the existence of a massive iron core: it must consist mostly of hydrogen. There are independent theoretical reasons for supposing that *all* materials become electrically conducting when sufficiently highly compressed, so it is assumed that liquid metallic hydrogen provides the first of the components essential for the generation of Jupiter's magnetic field.

But what about the second, the source of energy? It has been known for a long time that, while the outer surface of Jupiter, the visible part, is extremely cold at minus 144 °C, the planet radiates out two and a half times more heat than it receives from the Sun. It is tempting to suppose that Jupiter might contain an abundance of radioactive elements whose decay generates this excess heat – after all, this is how the Earth derives its own internal heat.

But radioactive elements are unlikely to be associated with the hydrogen and helium making up most of the mass of Jupiter. The most likely suggestion seems to be that the heat radiated from Jupiter is simply its original heat, trapped within it at the time of its formation. Unlike the Earth, which can maintain its internal temperature from radiogenic heat, Jupiter must be slowly and steadily cooling. A consequence of this is that the planet is actually *shrinking*, albeit at an immeasurably slow rate.

There is one other complicating factor. The rate at which Jupiter radiates heat is much greater than one would expect from a simple, solid body. Such a body would cool from the outside inwards, and heat would only be supplied slowly to the outside by *conduction* from within. The much greater *observed* rate of heat flow is achieved by *convection*. 'Hot', relatively low-density liquid hydrogen probably rises upwards through the body of Jupiter, carrying heat with it; it cools near the surface, becomes denser and descends once again.

Such convection currents, taking place within the liquid metallic hydrogen part of Jupiter, would also provide an elegant source of energy to generate the magnetic field; they are analogues of the circulating currents in the liquid iron of the Earth's core.

The result of all these lines of evidence is a hypothetical internal structure for Jupiter, such as that shown in Figure 8.2. There is a small rocky core, forming the concentration of mass at the centre, surrounded by a thick envelope of liquid metallic hydrogen which

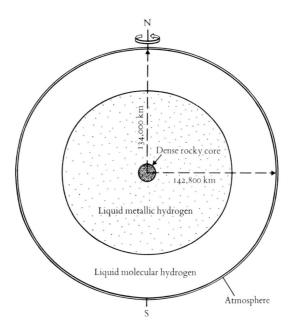

Figure 8.2.   The internal structure of Jupiter.

is constantly stirred up by convection cells, and this is bathed in an enormously deep 'ocean' of ordinary molecular liquid hydrogen. Above this is the atmosphere, the one part of Jupiter that can be observed directly.

It is important to emphasize that this structure *is* hypothetical, and that many plausible variations exist. One possibility is that there might be a great deal of water present, in place of hydrogen. Another is that there might be substantial quantities of rocky material thinly dispersed in the outer envelope. There is a great deal more to be learned about Jupiter, such as the way in which temperature varies with depth, before its structure can be confidently delineated. NASA's Galileo mission, scheduled for launch in 1982, is designed to drop a probe through the atmosphere: it will add dramatically to our knowledge.

Jupiter's atmosphere

In terms of volume, Jupiter's atmosphere is a minor part of the planet, but it is easy to observe. Telescopic observations since the seventeenth century have enabled the shifting belts and patches of cloud to be minutely documented, and these early observations are of immense value when interpreting the stunning close-up views obtained by passing spacecraft.

Spectroscopic studies over the decades have given progressively better data on the composition of the atmosphere. Hydrogen, helium, ammonia, methane and water are all present. It seems as though the light- and dark-coloured bands which characterize the surface are caused by warm, upwelling currents which cool as they rise, forming clouds of ammonia crystals suspended in gaseous hydrogen, just as terrestrial clouds are produced by the condensation of water vapour droplets on rising currents of air. The cooled gases descend in the dark bands between cloud stripes in a fairly regular way. Superimposed on this convection process, however, are the tremendous Coriolis forces caused by the planet's rapid rotation. These cause gas which would have moved towards the equator to swerve into a direction *opposite* to the rotation

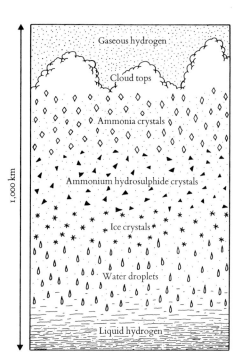

Gaseous hydrogen

Cloud tops

Ammonia crystals

Ammonium hydrosulphide crystals

Ice crystals

Water droplets

Liquid hydrogen

1,000 km

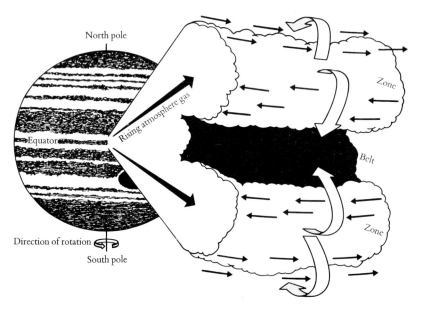

North pole

Zone

Rising atmosphere gas

Equator

Belt

Zone

Direction of rotation

South pole

Figure 8.3. (top) *One suggestion for the structure of Jupiter's atmosphere. The diagram is entirely hypothetical – direct observations only provide information about the topmost cloud layers.*

Figure 8.4. (bottom) *The conspicuous light and dark bands in Jupiter's atmosphere are created by ascending (light) and descending (dark) plumes of gas. Coriolis forces also cause the gases flowing from the poles to form winds blowing counter to the rotation direction; gases flowing towards the poles form winds blowing in the same direction as the rotation.*

direction, while gas heading for the poles ends up moving around Jupiter *parallel* to the rotation direction.

The result of these conflicting forces is that there are extremely powerful winds blowing in Jupiter's upper atmosphere, and their effects are evident in the swirling, turbulent cloud patterns. Jet streams have been detected which are more than 80,000 kilometres long (many times the diameter of the Earth), in which the wind velocities may be as much as 350 kilometres per hour. Tightly spiralling cyclonic structures are also clearly visible, resembling terrestrial revolving storms, but so large that a single one could swallow the Earth.

So much for the overall appearance. The details are much more complex. For one thing, the bands are not simply light and dark; there are many subtle shades of browns, yellows and reds. These cannot be explained in terms of the simple gases detected. Many more complicated chemical groups may be present. Among them may be ammonium hydrosulphide, various organic compounds and complex inorganic polymers. It has even been suggested that the brownish coloration could be an indication of conditions favourable for the evolution of life. All the necessary ingredients are present: light, warmth (at a sufficient depth) and the right kinds of organic molecules.

That great science-fiction writer, Arthur C. Clarke, described in one of his most brilliant stories, 'Meeting with Medusa', the exploration of the upper reaches of Jupiter's atmosphere by an astronaut in a 'hot hydrogen' balloon. The astronaut encountered

Figures 8.5 and 8.6.   Voyager 1 revealed many details of Jupiter's atmospheric structure which had never been seen before. Complex swirls and eddies abound in the composite photograph. Figure 8.6 summarizes some of the complex atmospheric circulation around the Great Red Spot. The white

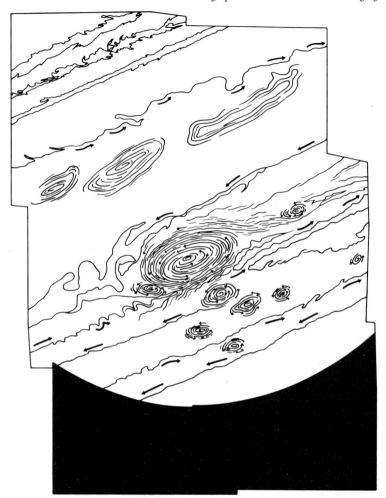

circular patches below the spot are believed to be enormous convection cells, where warm gases (pale tones) rise in a spiral fashion at the centre and turn downwards to form the outer, darker rings. The Great Red Spot and most of its smaller spiralling companions all have an anticlockwise rotation sense.

all manner of low-density organisms living suspended in the thick atmosphere like jellyfish in terrestrial seas. While this account was purely fictional, it is a wonderful stimulus for the imagination – this writer, in particular, can hardly wait for the results of the Galileo mission! Will it reveal giant floating organisms, as tall as the Empire State Building, drifting serenely through the Jovian atmosphere and quietly metabolizing methane?

The Great Red Spot

While the existence of life within Jupiter remains fictional, there is no longer much room for speculation about the nature of the enigmatic Red Spot. Thought to have been first observed by Robert Hooke in 1664, the spot has been persistently visible ever since, but has varied in size, colour and location. Its width is fairly steady at about 14,000 kilometres, but its length varies between 30,000 and 40,000 kilometres over every few years. Its colour also varies from a striking brick-red to a barely perceptible greyish pink. Most puzzling, however, is the way that it moves around: it shifts by several thousands of kilometres from side to side over the years, in a quite irregular manner.

These unpredictable variations, of course, presented difficulties when trying to explain what the spot is. The earliest ideas were that it was a great volcano, rearing its head high above the clouds. Although attractive, this idea does not fit in with the structure of Jupiter. And, of course, a volcano should not go skating around on the surface of a planet – it should stay in one place.

A more sophisticated variant of this idea argued that the spot was a giant cloud formation, generated by a standing wave downwind from a major mountain on the surface of Jupiter. This has the advantage that the position of the standing wave could shift with the strength of the wind, and that the suggestion that the spot was an atmospheric disturbance was a step in the right direction. The present thinking is that, in the absence of any solid surface, the spot is merely a persistent cyclonic disturbance, a bit like a giant hurricane. Hurricanes on Earth generally blow themselves out after a few days – it would be rather inconvenient if they lasted for three hundred years – but if they could be kept in one place, and a steady supply of energy were available, there is no reason why they should not last indefinitely. (Terrestrial hurricanes run out of steam when they move from warm, moist oceans on to dry land or cooler water masses.) The Great Red Spot, for all its side-to-side movements, appears to be confined to a particular latitude on Jupiter, and may be steadily refuelled from an atmospheric energy source. Furthermore, the spot is so massive, and contains so much momentum, that it could *not* simply blow itself out overnight – it is *bound* to keep going.

The satellites of Jupiter

Jupiter's family of satellites has played an important role in astronomy in several ways. The four biggest – Io, Europa, Ganymede and Callisto – are known as the *Galilean* satellites, because they were discovered by Galileo with his first telescope in 1609. The crudest form of optical aid will show them up easily; some lynx-eyed people claim to be able to see them with the naked eye. Galileo's observation was of much greater significance than the mere discovery of a few distant specks of light. He showed conclusively for the first time that objects in the solar system do not all rotate around the Earth, or even the Sun. This was a definitive nail in the coffin of the old Earth-centred school of thought.

In the centuries since Galileo, a further ten satellites have been discovered, bringing the total to fourteen. The thirteenth was discovered telescopically in 1974; the fourteenth by the Voyager 2 spacecraft in 1979. There has been rather an acrimonious debate in astronomical circles about the desirability of naming the nine satellites. The appropriate committee of the International Astronomical Union has proposed rather obscure mythological

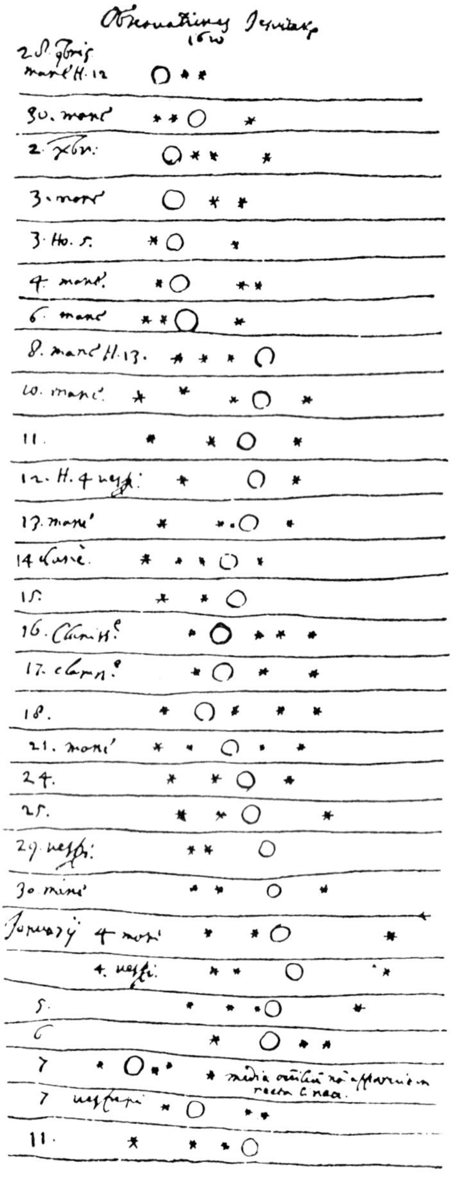

308

Figure 8.7. Galileo's meticulous records of his first observations of the relative positions of Jupiter's four large satellites, made in December 1609 and January 1610. Note that he records details of the 'seeing' conditions (e.g. 16 December 'clarissime') and that he made an observation on Christmas Day 1609.

names for them, such as Leda, Lysithea, Ananke and Pasiphae. These proposals have not gone down well with some of the observers most intimately concerned, such as Edward Barnard and Charles Kowal, who both discovered satellites. They prefer a much more practical approach: each satellite should be given a number. Thus Pasiphae would become JVIII, and Ananke JXII. In this way, each satellite in the solar system could be uniquely identified. Not so attractive, perhaps, but definitely practical.

Long before the dimmer satellites of Jupiter had been discovered, and before the quarrels over their names had erupted, the Galilean satellites had helped science to make a great advance. One of the most important statistics in the whole of science is the velocity of light. This value is now known to a high degree of precision, but it was Jupiter's satellites, and Io in particular, that first enabled this measurement to be made *three centuries ago*. A brilliant Danish astronomer, Ole Rømer, a colleague at the Paris observatory of the great Giovanni Cassini, became interested in the times at which Io was eclipsed by Jupiter.

Since Io revolves around Jupiter in a near circular orbit, it should pass behind Jupiter (as seen from the Earth) in a perfectly predictable fashion, a fixed number of hours, minutes and seconds apart. Rømer found that although there was a regular period (about forty-two hours) between one eclipse and the next, he could not *predict* when eclipses would take place a matter of months ahead. Sometimes they seemed to be taking place before the time he predicted, sometimes afterwards. He realized that eclipses were taking place ahead of schedule when the Earth and Jupiter were close together in their orbits, and behind schedule when they were far apart.

Rømer had the brilliant insight to see that the errors in his predictions were caused by *the time taken by light* to cross the space between Jupiter and Earth. When the distance was least, the time taken was least, and Io therefore *seemed* to disappear behind Jupiter 'early'; when the distance was greatest, the time taken was greatest, and Io therefore seemed to disappear 'late'. At the extreme, there was a difference of over sixteen minutes between predicted and actual times. Rømer was able to deduce from these figures that the velocity of light was about 200,000 km sec^{-1}. (The modern figure is 300,000 km sec^{-1}.) This measurement,

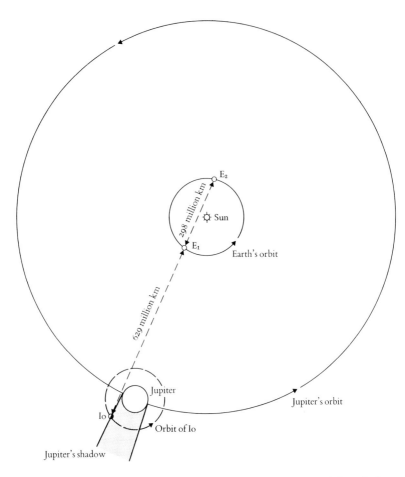

Figure 8.8.   The principle underlying Romer's measurement of the velocity of light. The distance from Earth to Jupiter is 298 million kilometres greater at E_2 than at E_1. It takes light about sixteen minutes to travel this distance, so Io appears to be eclipsed earlier when seen from E_1 than E_2.

made with extremely simple instruments, paved the way for some of the most fundamental modern work in physics and astronomy. The best-known equation of all time is Einstein's $E = mc^2$, c being the velocity of light.

The Galilean satellites

Much the most important of Jupiter's fourteen satellites are the innermost five, consisting of Amalthea, about 150 kilometres in diameter, and the four big Galilean satellites. These really are *big*. Two, Ganymede and Callisto, with diameters of 5,270 and 5,000 kilometres respectively, are bigger than the planet Mercury. Io (3,640 kilometres) and Europa (3,100 kilometres) are much the same size as the Earth's Moon (3,476 kilometres).

The densities of the Big Four are particularly informative. There is a steady decrease outwards from Io ($3 \cdot 6 \times 10^3$ kg m^{-3}) to Callisto ($1 \cdot 7 \times 10^3$ kg m^{-3}). Io and Europa both have densities close to that of the Earth's Moon; it can therefore be safely assumed that they are made of the same kind of rocky silicate material. Ganymede and Callisto, by contrast, must contain a large proportion of low-density material, possibly ordinary ice, or other condensed gases. All four satellites are locked into synchronous orbit around Jupiter; like the Earth's Moon and Mars' Phobos and Deimos, they always keep the same face turned towards their parent planet.

Their orbital periods also exhibit some remarkably elegant resonances. The orbital period of Io is $1 \cdot 77$ days, of Europa $3 \cdot 55$ days, of Ganymede $7 \cdot 15$ days and of Callisto $16 \cdot 69$ days. Europa's period is twice that of Io, and Ganymede's is twice that of Europa! This is no mere coincidence; the resonances between the satellites are dictated by their complex gravitational interactions.

It is clear from their sizes and densities that the big satellites are in effect really planets in their own right which are overshadowed by their enormous parent. They are all big enough to show distinct discs when seen through major telescopes, and some surface markings can be made out on them. An enormous

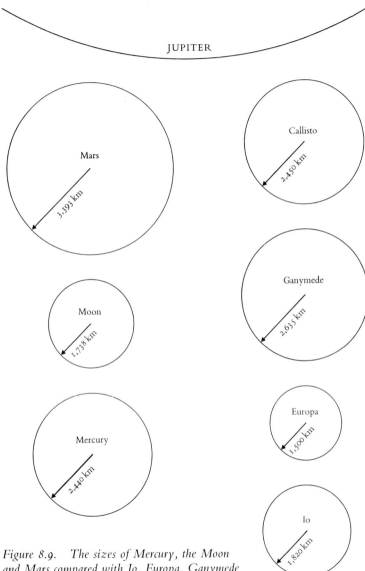

Figure 8.9.   The sizes of Mercury, the Moon and Mars compared with Io, Europa, Ganymede and Callisto, and Jupiter.

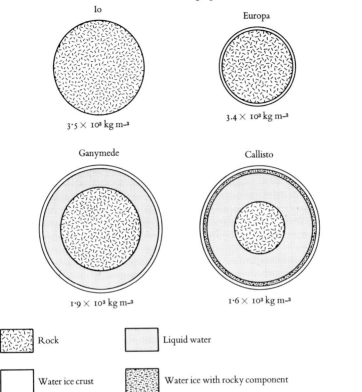

Figure 8.10. *Possible internal structures of the four Galilean satellites.*

amount has been learned about the satellites in the last few years, not only through spacecraft missions, but also through meticulous telescopic observations, some of them made from telescopes carried in high-flying aircraft.

Io

As early as the turn of the century, Io had been recognized as being something of an oddity. Edward Barnard at the Yerkes Observatory reported that it had *red* polar caps. Now, polar caps in themselves are no surprise, but *red* ones are more difficult to explain. Later – and most unusual for a satellite – it was discovered that Io has an atmospheric pressure of one twenty thousandth that of the Earth's. And, in 1974, Robert Brown of Harvard University discovered that Io is surrounded by an orange glow like that from a sodium street lamp. This peculiar aura is caused by the scattering of sunlight by sodium atoms which have been sputtered into space from the surface of the satellite.

But where could the sodium come from? Careful infra-red telescopic studies of Io showed that its surface is almost completely free of water ice and that its pronounced reddish colour and infra-red spectral characteristic could best be explained by deposits of ferrous salts on its surface and quantities of nitrate salts. Some sodium salts may also be present. All these salts were thought to be evaporites, deposited from solutions leached out by water deep beneath the surface and brought up to the surface.

In a particularly brilliant piece of scientific hypothesizing, it was also predicted that the heat generated by the powerful tidal disruptions within Io's interior, caused by its proximity to the enormous mass of Jupiter (only 421,000 kilometres away) and by its resonance with Europa, would be enough to cause melting of its interior and hence that widespread volcanism would be visible at its surface.

Such speculations naturally meant that Io was a prime target for spacecraft investigation, but, of course, no single mission could be devoted to one satellite. Voyager 1 passed right through Jupiter's satellite system, penetrating between Jupiter and Io's orbit, and had close encounters with both Io and Ganymede, but came much less near to Europa and Callisto. Voyager 2 passed close to Europa, Ganymede and Callisto, but not, unfortunately, to Io.

The pictures of Io obtained by Voyager 1 were among the most scientifically rewarding of all the many thousands of planetary pictures to be obtained over the last decade. Although Io is roughly the same size and density as the Moon, the Voyager pictures showed that it is quite different. The Moon's surface is ancient, and covered by myriads of impact craters. Io's surface is young, lacks obvious impact craters but *does* have many volcanic centres, with lava flows radiating out from them. One particularly amazing picture also showed a great plume of material being blown off into space. Clearly, there are *active* volcanoes on Io, exactly as the calculations of tidally induced heating had suggested. No less than *seven* separate active volcanoes were located on the Voyager 1 pictures.

The confirmation that volcanism is actively under way on Io raises some immensely exciting questions for geologists, which will take many years – and perhaps even landing missions – to answer. What composition are the lavas? Are they basalts like

Figure 8.11. One of Voyager 1's historic pictures of Io. The surface is reddish in colour, and has few impact craters. Many volcanoes are visible, some of them surrounded by radiating lava flows. The pale-coloured areas may be patches of frost.

those on the Earth and Moon? What is the internal structure? And what kind of surface processes are active? A key problem is the nature of the volatile material that 'drives' the volcanic activity. On Earth, this is water, which is blown off in the form of steam. Io appears to lack water, so its volcanism may be driven by *sulphur* which is somehow recycled from the crust back into the interior.

Figure 8.12.   One of the most extraordinary photographs ever taken, this view of Io shows two volcanic eruptions in process. One is seen on the limb of the satellite, spraying ash clouds over 200 kilometres into space above the surface, before falling back along parabolic ballistic paths. The second is on the boundary between light and dark areas, and is seen as a bright patch because it is high enough to catch the rays of the rising sun.

There is no doubt that sulphur is extremely abundant on the surface of Io, and some scientists have suggested that the lava flows observed on the surface may be composed of pure sulphur, rather than rocky materials. Although rare, sulphur lava flows are known on Earth, so this suggestion is reasonable enough. Some of the other ideas on the geology of Io have more of the flavour of science fiction than science about them, although they are undoubtedly realistic.

Consider, for example, the crust of Io. One suggestion is that there is a deep layer of liquid silicate material (representing the 'mantle'), overlain by a crust of solid silicates which has sufficient topography to rear up in places to form 'islands'. Resting on low-lying parts of the crust and surrounding the silicate 'islands' is an 'ocean' of molten sulphur, which is overlain in turn by an outer skin of solid sulphur and sulphur dioxide. A further refinement of this structure suggests that just below the surface of the outer skin, pressure and temperature conditions will be such that sulphur dioxide can exist as a *liquid* which behaves like water in a terrestrial artesian basin! Whenever the skin is ruptured for some reason, the liquid sulphur dioxide surges out in a flood, sapping and carving away at the solid parts of the skin, and leaving erosional scarps which are characteristic features of Io's topography.

Apart from its role as a most unusual agent of erosion, the escaping liquid sulphur dioxide also accounts for the huge volcanic plumes seen on Io. As soon as it breaks surface and encounters the near-vacuum of Io's atmosphere, the liquid sulphur dioxide volatilizes explosively, blowing off as great plumes 100 kilometres or more into space. The gaseous sulphur dioxide then rapidly condenses in the chill of space to the solid form, and showers back to the surface in the form of tiny 'ice' particles.

The continuous spewing of sulphur dioxide into space, and its return as a 'snow' of solid particles, mean that the whole of Io is being constantly resurfaced, and accounts for the lack of impact structures on the satellite. One calculation suggests that material is added to the surface at a rate of about 0·1 centimetres per year, extraordinarily rapid by terrestrial standards. Such 'resurfacing' does take place on Earth, however. Almost all the oceanic crust has been created by volcanic activity in the last 100 million years.

Speculations such as these mean that Io is clearly going to require much further investigation before it can be satisfactorily understood. Like every other space mission Voyager raised more questions than it answered!

Europa

Like Io, Europa is about the same size and density as our Moon. Earth-based telescopic studies showed that it is an extremely bright object, and that almost 90 per cent of its surface area is

Figure 8.13.   Europa looks almost as smooth as a billiard ball in this rather distant Voyager 1 view, apart from a few long linear structures. The smoothness of its surface is probably because of a constantly renewed coating of fresh ice.

covered with water ice. The Voyager pictures revealed a rather uniform, smooth-looking sphere, its bland surface broken only by a few large linear features. This suggests that the surface of Europa is indeed an almost unbroken sheet of ice, and furthermore, because there is little evidence of *any* cratering, that the surface must be renewed from time to time from within.

Although Europa is further away from Jupiter than Io, it is still near enough for some tidally induced heating to occur in its interior. It must also contain within it a small quantity of heat-producing radioactive elements. The heat produced by these may be sufficient to cause melting of the lower parts of the ice cover, and for the eruption of watery 'lava' flows at the surface. These would spread out over the surface, smoothing out irregularities, before freezing into ice.

Why should there be so much water (ice) on Europa, and so little on Io? The answer probably is that volcanism has been so intense on Io for such a long time that all the water originally present has long since been driven off. Steam is much lighter than sulphur dioxide, and thus its chances of escaping into space, rather than being recycled to the surface, are about ten times greater than those of sulphur dioxide.

Ganymede

Voyager 1 passed only 250,000 kilometres from Ganymede, and obtained some superb pictures. A smooth, bluey-grey sphere pock-marked with craters, Ganymede looks superficially much more like our Moon than either Io or Europa. In the close-up views, features as small as 4 kilometres in diameter are visible, which is comparable with what can be seen on the Moon with a modest telescope.

Earth-based telescopic studies of Ganymede had suggested that just over half its surface would be covered by water ice, much less than Europa. This may seem surprising at first sight, especially since Ganymede is strikingly less dense than Europa and must contain a larger amount of water in its interior. The anomaly is explained by the fact that Ganymede's surface is partially covered with dust, derived from the small and irregular outer satellites of Jupiter. This dust arrives on the surface of Ganymede at relatively low velocity, unlike the micrometeorites that bombard the Moon's surface, and so coats the surface like coal dust on a miner's face, rather than blasting into it and volatilizing it away.

The Voyager pictures could not, of course, resolve this fine dust, but they did clearly show fresh, young craters on Ganymede caused by large impacts which flung streaks or rays of brilliant

Figure 8.14.   Ganymede, a satellite of Jupiter which is as large as Mercury. The bright patches are impact craters, like those on the Moon, but the surface is probably dust-covered ice rather than solid rock.

white ejecta for hundreds of kilometres from the impact site. The ejecta is thought to be clean, fresh ice, but, interestingly enough, the surface over which it is spread is much darker, suggesting that it is indeed covered with a film of dark dust.

It is also clear that Ganymede has experienced a similar kind of watery volcanism to Europa. Although the surface *is* cratered, there are far fewer craters than on the Moon, and the surface has clearly been renewed sufficiently to smooth out most early-formed craters.

Figure 8.15. Bright streaks or rays of ejecta from impact craters are clear in this view of Ganymede. The small number of craters indicates that the surface has been renewed from time to time.

Callisto

Ganymede has less ice on its surface than Europa, notwithstanding that it has much more ice in its interior. Similarly, Callisto, which is less dense even than Ganymede, and must be made largely of ice, has almost no surface water ice. Since Callisto is further out from Jupiter than Ganymede, it seems to have received a much heavier coating of dust; so heavy that its entire surface is covered. The dust also explains why Callisto is much the darkest of the Galilean satellites – its albedo is only 0·2. The Voyager spacecraft only obtained rather distant views of Callisto, but these revealed that it is a heavily cratered body. It boasts one spectacular impact basin, surrounded by ten beautifully concentric rings. This basin was clearly produced by a major impact, and its presence suggests that the surface is ancient, and that Callisto has *not* experienced the smoothing effects of watery volcanism that Europa and Ganymede show.

Figure 8.16.   Callisto appears to be much more heavily cratered than Ganymede, but the reasons for this are not understood.

Clearly, Jupiter's four Galilean satellites are very different. Each is a world of its own, with its own geography and geological history. A great deal remains to be learned about these distant bodies which is bound to help us to understand the evolution of our own small planet and its solitary attendant. Many of the biggest problems will only be resolved by landing spacecraft on the satellites. Planetary scientists all over the world are already day-dreaming about possible missions to Io and Ganymede. Technic-ally, it would not be much more difficult for an astronaut to land on Ganymede than on our Moon . . .

Jupiter's other satellites

The unusual properties of the large Galilean satellites tend to distract attention from Jupiter's retinue of lesser attendants. Little is known of these, except that they are small and insignificant.

Voyager 1 passed within 415,000 kilometres of Amalthea, the innermost satellite, which orbits about 160,000 kilometres from Jupiter. Amalthea seems to be a rather odd body. Its shape is extremely irregular – it is twice as long as it is wide – and its long dimension of only 290 kilometres always points towards its parent planet. The Voyager pictures revealed it to be extremely dark, with a puzzlingly low albedo and a number of crater-like features on its surface.

Much more surprising than these views of Amalthea, however, was the discovery of a *ring* around Jupiter. The ring appears to be made up of widely spaced boulders and rock debris, and is about 30 kilometres thick, 80 kilometres broad and encircles Jupiter about 50,000 kilometres above the cloud tops. The presence of the ring was totally unexpected, and could never have been seen from Earth.

Figure 8.17.   Amalthea, Jupiter's innermost satellite, was snapped by Voyager 1, and was found to be a very irregularly shaped body, probably heavily cratered like Phobos and Deimos.

Of the other satellites, one group of four, ranging up to 160 kilometres in diameter, orbits Jupiter at a distance of about 11 million kilometres. The remainder form a second distinct group, all of them less than 20 kilometres in diameter, which orbit the planet at a distance of about 22 million kilometres in a *retrograde* sense; that is, they go round the planet the opposite way to the other nine. While this group of satellites may represent a cluster of captured asteroids, the mechanism whereby they ended up going the wrong way around the planet is not understood. The tight clustering together of the two groups has also been used to argue that they represent the shattered fragments left when two older, larger moons of Jupiter were destroyed by collisions with large asteroids. Massive collisions were really rather common events in the early history of the solar system, as the final chapter will emphasize.

9. Saturn and the Outer Planets

Saturn

Saturn is a popular planet. Not only is it the single planet that has a day of the week named after it – the English *Saturday* – but it is also the only planet that most people could unhesitatingly draw a sketch of that would be recognizable. The feature that makes Saturn so unmistakable is its system of rings. These are as intimately associated with Saturn as canals are with Mars, the difference being that the Saturn rings really do exist.

Figure 9.1.   This picture of Saturn was obtained by Stephen M. Larson on 11 March 1974 at the Catalina Observatory, and is a composite of sixteen separate images. The important components of the rings are well displayed: the fuzzy outermost A ring, the dark Cassini Division, the bright, solid-looking B ring, and the fuzzy crêpe or C ring. Light and dark banding on the planet's surface is also visible.

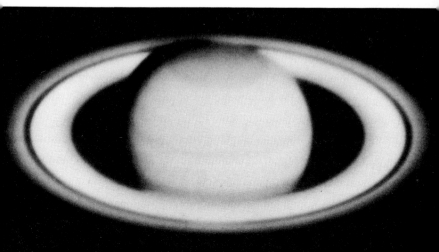

Although the elegant symmetry of its rings makes Saturn the most strikingly beautiful member of the solar system, the rings are a minor part of the planet: they are vanishingly thin, and contain only a tiny fraction of the planet's mass. So while the rings richly deserve description, and will get their fair share here, it is important to keep them in perspective: they are merely decorative appendages to the planet proper, as inessential to it as the clothes draped on the body of a beautiful woman, and almost as insubstantial as the halo of a saint.

Saturn's statistics

Saturn orbits the Sun at a mean distance of 1,427 million kilometres, twice as far out as Jupiter, and is the last of the planets to be easily visible to the naked eye. The equatorial diameter is 120,000 kilometres, but the polar diameter is *10 per cent* less than this, making Saturn much the most flattened of the planets.

Because of its immense distance from the Earth, Saturn is not a rewarding object for telescopic study, apart from the visual splendour of its rings. The body of the planet appears yellowish, and there are a number of bands of different colours running parallel to the Equator. These are similar to the stripes on Jupiter, but much less detail can be made out. As with Jupiter, the bands nearer the poles rotate more slowly than those near the equator, demonstrating that they are atmospheric manifestations, not parts of the solid surface of the planet. Some faint light and dark mottlings are visible, and from time to time conspicuous bright patches appear, but there is nothing to compare with Jupiter's Red Spot. At the time of writing, only the Pioneer 11 spacecraft had reached Saturn. Its low resolution pictures did not add a great deal to our knowledge of the surface appearance of Saturn.

There is not much that can be said either about the internal structure and composition; in both these respects, Saturn appears to be broadly similar to Jupiter. Spectroscopic studies of the atmosphere have demonstrated the presence of hydrogen and methane, but there seems to be much less ammonia than on Jupiter. Gases such as ethane, ethylene and acetylene may also be present in small quantities, since these could be produced by the effects of solar radiation on methane in the atmosphere. These organic chemicals might be present as a 'smog' of small, dark particles which might warm up the atmosphere by absorbing sunlight and transferring heat to the surrounding gas. The methane, too, might cause atmospheric warming through a 'greenhouse' effect.

Saturn's density of only 0.7×10^3 kg m^{-3} is similar to that of balsa wood, much lower even than that of Jupiter. There probably is not as much difference between the internal structures of the two planets as the density contrast suggests. Like Jupiter, Saturn also probably consists predominantly of hydrogen in liquid and metallic forms, with a tiny rocky core. The fact that Saturn is so markedly flattened suggests that the dense core of Saturn probably contains a smaller proportion of the planet's total mass than does Jupiter's.

The Rings

The notion that Saturn has rings is so familiar to us in the twentieth century that it may come as a slight surprise to learn that it took the early astronomers fifty years to realize exactly what they were. Galileo – of course – was the first to observe Saturn, but although he noticed that the shape of the planet looked a bit odd, he could not be more specific. Later on, seventeenth-century astronomers, straining their eyes to interpret the distorted images in their telescopes, drew all kinds of strange configurations. Some showed large lumps attached to the side of the planet like ears, others showed structures like a pair of handles attached to a soup bowl, while still others drew three quite separate bodies. It was not until 1655 that the great Dutch astronomer, Christian Huygens, after years of observation, showed that the 'lumps' and 'handles' were in fact a thin disc surrounding the planet, but touching it nowhere.

It was realized as early as the eighteenth century that the rings could not be solid discs. By 1850, E. Roche had demonstrated from theoretical principles that any solid body that came within about 2.4 times the radius of the planet would simply be shattered

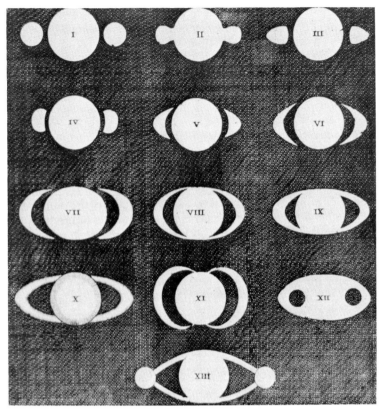

Figure 9.2.   Some sketches of early telescopic views of Saturn and its rings. At top left is Galileo's telescopic interpretation of the rings. From Huygen's Systema Saturnium *(1659).*

by gravitational forces, and the fragments would be dispersed into a ring around the planet. James Clerk Maxwell arrived independently at the same conclusion: the rings can only consist of myriads of tiny particles orbiting around the planet at different speeds and distances in accordance with Kepler's laws. The conclusion is reinforced by detailed study of the rings with large telescopes. Three separate rings are visible, and the inner and

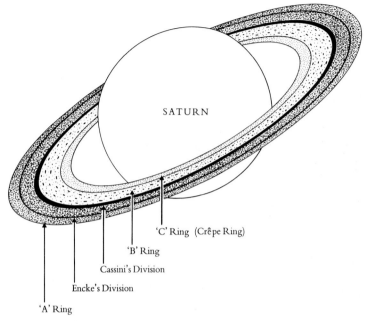

'C' Ring (Crêpe Ring)

'B' Ring

Cassini's Division

Encke's Division

'A' Ring

Figure 9.3. The detailed structures of Saturn's rings.

outermost ones are so insubstantial that stars can be seen through them on occasions. The innermost ring is known as the crêpe ring, on account of its gauzy, diaphanous nature.

Saturn's rings present a magnificent sight because they are usually inclined at an angle to the Earth, so that we can see a broad stretch of the plane of the rings. Every fourteen years and nine months, however, the geometrical relationships between Earth and Saturn are such that the Earth lies exactly in the same plane as the rings. When this happens the rings seem to disappear completely!

This indicates, of course, that the rings are extremely thin. Estimates of the thickness have gradually shrunk with the passage of time, and when the Earth last passed through the plane of the rings in 1966, it was estimated that the rings could not be more than 1·3 kilometres thick! The width of the rings is about 270,000 kilometres; hence the ratio of thickness to width is about 1:207,000. The ratio of thickness to width of an average page in a Pelican is about 1:2,7000. Thus, Saturn's rings are nearly a hundred times thinner proportionally than the sheet of paper you are reading!

The rings obviously cannot contain a large amount of material, but it is of considerable interest to know what it could be, and vigorous efforts have been devoted to finding out. Photometric, polarimetric, spectroscopic, radar and microwave studies have all been made, and they indicate that the particles are mostly ice, in lumps ranging from 30 to 40 centimetres. Rather than glossy ice cubes orbiting the planet, the particles probably have a texture like that of snow, and may be impregnated with 'dirt' in the form of iron sulphide minerals, with, possibly, a trace of silicate materials.

These conclusions are drawn from Earth-based observations. The rings are a natural target for spacecraft investigation at first hand. It has even been proposed that spacecraft should be sent right *through* the rings, or through one of the divisions between them. Recent estimates of the particle size distribution within the rings suggests that this might not be a very healthy thing to do, as there is only a 1 in 40,000 chance of the spacecraft *not* hitting anything!

Saturn's satellites

Saturn has eleven satellites. The first four of them were identified by Cassini in the seventeenth century; the eleventh was spotted as recently as 1979 by the Pioneer 11 spacecraft. All of them have been given rather esoteric – many would say confusing and redundant – names from Greek mythology, such as Mimas, Enceladus, Tethys, Dione, Rhea and Titan. Of these names, only *Titan* is at all helpful – Titan is a truly titanic satellite, some 5,800 kilometres in diameter, and much bigger than the planet Mercury (4,880 kilometres). The other satellites are all much smaller, most of them less than 1,000 kilometres across. It is difficult precisely to determine their masses, and hence their densities, but the densities appear to be low, indicating that the satellites are probably composed of ice and frozen gases.

In 1978, claims were made that at least one other satellite had been discovered between the rings and the first of the well-established satellites. Although this claim has not been validated, there seems good reason to suppose that there may be several small satellites orbiting Saturn just beyond its ring system. They may, in fact, constitute parts of an extended, discontinuous ring themselves.

Eleven satellites are too many to discuss individually, but the orbital periods of Saturn's satellites contain some of the most beautiful statistics in the solar system, so it is worth summarizing them as in Table 9.1. Careful inspection of the period column will

Satellite	Orbit radius (distance from Saturn) km	Period (days)
Janus	158,500	0·748
Mimas	186,000	0·942
Enceladus	238,000	1·370
Tethys	295,000	1·888
Dione	377,000	2·737
Rhea	527,000	4·518
Titan	1,222,000	15·95
Hyperion	1,481,000	21·28
Iapetus	3,560,000	79·33
Phoebe	12,930,000	550·4

Table 9.1. The orbital periods of Saturn's satellites.

reveal some wonderfully harmonious relationships. Mimas orbits around Saturn *twice* in exactly the same time that Tethys takes to go round *once*. Enceladus also revolves twice as fast as Dione, while Titan completes *four* revolutions in the same time that Hyperion takes to complete *three*.

The origin of these elegant resonances between the satellite orbits is not fully understood, but it is almost certainly connected with the kind of tidal interactions between satellites and their parent planets that are responsible for our Moon's gradually moving away from the Earth and revolving more slowly around it. In the case of Titan and Hyperion, their resonance might be an original feature, dating back to the time when Saturn and its satellites first formed. In that case, it is likely that Hyperion will show the same kind of spin-orbit coupling as Mercury – namely, rotating *three* times on its axis while orbiting Saturn *twice*. Saturn is far too distant for it to be possible to check this by telescopic observation, so it will require a spacecraft to confirm that this complex harmony exists.

Titan

Titan is worthy of a special mention, and not only for its sheer size. Its density of 1.4×10^3 kg m^{-3} suggests that it may have a small rocky 'core', surrounded by a 'mantle' consisting mostly of water with small amounts of rocky debris and with, perhaps, some ammonia dissolved in the water. The 'crust' may be a mixture of water, ice and methane.

This structure suggests some bizarre possibilities for the 'geology' of Titan. If the surface temperature is as high as some scientists think, then the ice in the crust may be present as water, and then there might be a layer of liquid methane floating on the water–ammonia solution. If the ice is only partially melted, there might be great 'continents' of ice drifting around, perhaps with 'volcanoes' erupting liquid methane. The possibilities are endless.

All this is pure conjecture. The atmosphere of Titan has been an even more fruitful source of conjecture. There is no doubt at all that Titan does have an atmosphere – it is the only planetary satellite to have a significant one. It was first detected in 1944 by Gerard Kuiper, who obtained conclusive spectroscopic evidence for the presence of methane. Subsequent work showed that the atmospheric pressure on Titan could well be much higher than that on Mars, might even be as high as that on Earth, and that other gases such as ammonia might be present. The excitement about Titan's atmosphere centres on the fact that it may be similar in some respects to the Earth's primordial atmosphere, and that it could therefore provide an environment in which life might evolve.

There are two main reasons for thinking this. First, all the right chemical elements are there: carbon, hydrogen, nitrogen and so on. Secondly, although Titan is extremely distant from the Sun, the methane in its atmosphere could act – like carbon dioxide – as a 'greenhouse', and allow temperatures to rise to viable levels.

There is no consensus of views on just what surface conditions on Titan are like. Some scientists argue that the greenhouse effect would not be anything like sufficient to bring temperatures up to the minimum at which metabolic processes could operate; others, that a large greenhouse effect would also result in the existence of aqueous solutions of ammonia on the surface, at concentrations sufficient to kill all known terrestrial organisms. (There may, of course, be Titanic organisms that thrive on strong ammonia solutions!)

The most optimistic exo-biologist, Carl Sagan, has argued that the greenhouse effect may be sufficient to bring temperatures only up to -73 °C, well below that required for terrestrial organisms to metabolize, but that it is probable that organic compounds exist in large quantities. So, while conditions are appropriate for the first purely chemical steps towards the origin of life to be made, it seems unlikely that living organisms actually exist on Titan. It is faintly possible that there may be pools of liquid water on the surface, warmed by heat from the centre of the satellite, and that in these life would stand a much better chance of evolving.

Chiron: mini-planet or asteroid?

Until November 1977, it was generally accepted that only a yawning gap of empty space separated the orbits of Saturn and Uranus. But then an American astronomer, Charles Kowal, made a discovery that may have some important consequences for our concept of the nature of the outer parts of the solar system.

As with so many other scientific discoveries, Kowal was looking for something quite different when he stumbled across Chiron. He was actually studying photographic plates of asteroids (Chapter 10), taken with one of the great telescopes on Mount Palomar in California, and was particularly looking out for the so-called *Trojan asteroids*, which are tied in to the orbit of Jupiter. You may recall from Chapter 1 (page 22) that Kepler's laws of planetary motion dictate that the further out a planet is from the Sun, the longer it takes to go around the Sun. This means, of course, that a distant object will appear to move more slowly against the starry background than a near one.

Thus, when Kowal found on his photographic plates a tiny point of light which had drifted only three minutes of arc in twenty-five hours, he knew immediately that he was dealing with something quite different from the Trojan asteroids, which would appear to move three times as fast, and he knew that, whatever the body was, it must be nearly as far from the Sun as the planet Uranus.

His discovery naturally aroused a great deal of excitement in astronomical circles, and feverish observations were made at several observatories to refine details of the object's orbit and to try and find out more about it. The object rapidly acquired a name, Chiron, and was found to move round the Sun in a period of 50·7 years in a strongly elliptical orbit, much more elliptical than any of the major bodies in the solar system, except for a few asteroids. As Figure 9.4 shows, its orbit crosses that of Saturn and reaches nearly as far out as Uranus.

But what is Chiron? It must be extremely small, because it is a mere speck of light even in a giant telescope, and would have perceptible effects on the orbits of Saturn and Uranus if it had a large mass. The initial reports of the discovery of Chiron were accompanied by speculations about it being a new 'mini' planet. It seems much more likely, though, that it is more like an asteroid than a planet, and that it may even be the first known member of a second belt of asteroids, like that between Mars and Jupiter.

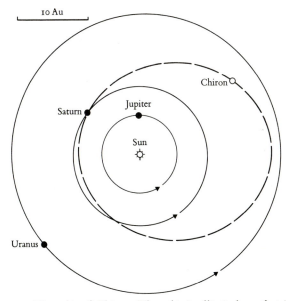

Figure 9.4. The orbit of Chiron. The orbit is elliptical, so that it crosses Saturn's orbit and reaches out nearly as far as Uranus.

Years of patient, meticulous observational work will be required to establish this point, but dedicated astronomers have already started on the long search for Chiron's companions.

Uranus

Uranus and the planets beyond are subjects of a different realm. All of the planets discussed so far were known to the ancients: with only the most rudimentary optical assistance the inquiring reader could see for himself that Jupiter really *does* have four large moons, strung out like beads on a wire, that Saturn *does* have rings, or that Mars *does* have polar caps.

Somehow, this makes these planets seem real, and accessible. But Uranus, Neptune and Pluto are so remote that they can only be studied with large telescopes, and even the largest scarcely shows Pluto as anything more than a pin-prick of light. Millions of people must have seen the rings of Saturn through a telescope; far fewer have ever gazed directly on Uranus; while the number that have seen Neptune or Pluto must be exceedingly small.

Because they are so inaccessible, it is easy to think of these planets simply as solar system statistics rather than as real worlds in space. Individually, though, they are just as significant as any of the other planets, and will undoubtedly figure much more importantly in discussions of the solar system when spacecraft make more information available.

Uranus was discovered on the evening of 13 March 1781 by the English astronomer, William Herschel. His discovery was accidental, in that he happened to notice a greenish 'star' in the constellation Gemini, which, according to the star charts, should not have been there. At the time, Herschel dismissed his observation as a mere comet, and it was not until a year later that it was confirmed as a planet.

Although dim by comparison with the other planets, Uranus is just visible to the naked eye on clear, dark nights, and it had been noticed and recorded by a number of astronomers prior to Herschel, but none of them twigged that it might be a new planet.

Uranus orbits the Sun at a mean distance of 2,800 million kilometres. Because it is so remote, it moves slowly relative to the starry background, taking eighty-four years to go once round the Sun. This is a slightly depressing statistic, because it means that no astronomer will ever live to see Uranus in the same position twice, unless he is fortunate enough to get started in astronomy while still wet behind the ears. Furthermore, the planet is so remote that no features can be seen even with balloon-borne telescopes flown above the Earth's atmosphere.

Notwithstanding this, it is possible to glean a surprising amount about Uranus. First, it is fairly big, with a diameter of 51,800 kilometres, not quite half that of Saturn. Secondly, it has a density of about $1 \cdot 2 \times 10^3$ kg m^{-3}, greater than that of Saturn, but slightly less than Jupiter. Because of the relative differences in volumes involved, this means that Uranus may have a proportionately larger core of dense, rocky material than either Jupiter or Saturn.

Spectroscopic studies have shown that Uranus has a different atmosphere from Jupiter and Saturn; it has much more methane and much less hydrogen and helium, and apparently little ammonia. The temperature of the planet seems to be about −215 °C, which is just about the same as that of a chunk of dark material at the same distance from the Sun. This means that Uranus' temperature is only that which can be attributed to solar warming: it does not *appear* to have any source of internal heat. In this respect, it is different from the other planets.

Uranus is also an odd-one-out in another, more striking way. The other planets are all conventional in that their rotation axes are all steeply inclined to the plane of their orbits. Uranus' axis, however, is very nearly in *the plane of its orbit*. The rotation period is about eleven hours. It is difficult to explain how this state of affairs came about, and especially to reconcile it mechanically with the usual ideas for the origin of the solar system. Whatever the cause, the peculiar inclination leads to some even more peculiar consequences for the planet's 'seasons'.

Almost every point on the Uranian globe passes in succession through almost all possible positions relative to the Sun. At the winter and summer solstices, the Sun is nearly overhead at one

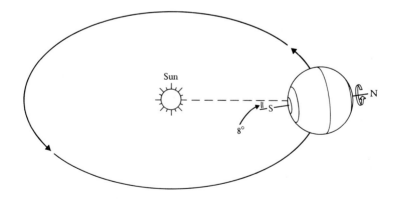

Figure 9.5.   Uranus is tipped over at such an extraordinary angle that its axial rotation is effectively retrograde.

of the poles, while the other endures decades of darkness. (At the terrestrial poles, the Sun never rises more than $23\frac{1}{2}$ degrees above the horizon.) The zones of perpetual sunlight (or darkness) also extend nearly as far as the equator. A Uranian living at the latitude of London or New York would experience about nineteen years of day and night, followed by twenty-three years of constant night, then another nineteen years of day and night, and finally twenty-three years of constant daylight!

Until 1977, the unorthodox way that Uranus trundled around its orbit was much the most interesting facet of the planet. On

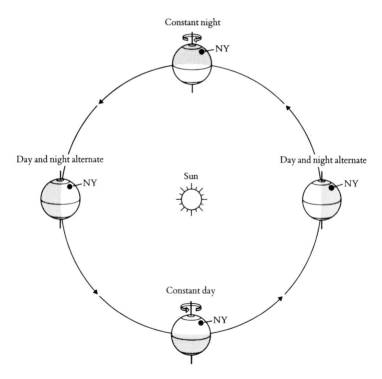

Figure 9.6.   The effect of Uranus' peculiar axial inclination on the sequence of day and night. NY denotes a point at the approximate latitude of New York on Earth.

10 March 1977, however, a totally unexpected discovery was made which abruptly focused attention on the planet. A carefully planned campaign had been organized to enable an extremely rare event to be observed: Uranus was due to occult, or pass in front of a star. The occultation was only visible from the southern part of the Earth – where there are few major observatories – so NASA dispatched its Kuiper Airborne Observatory to Australia to ensure that the event would be monitored. The airborne observatory (named after Gerard Kuiper) is exactly what its name implies: a high-flying jet aircraft carrying a computer-stabilized astronomical telescope.

The occultation was expected to take place at 21.05 hours. At 20.00 hours, the flying observatory was airborne at 41,000 feet over the Pacific, maintaining a steady course that would give it the best possible view of Uranus. On the ground at Perth, Western Australia, the 24-inch telescope at the Perth Observatory was also locked on to Uranus, so that simultaneous observations could be made. The aims of the joint exercise were to determine accurately the diameter and oblateness of the planet, and to investigate the temperature of Uranus' atmosphere. The main instrument to be used was a sensitive photometer attached to the telescope, to measure the intensity of the light coming from the star which Uranus was due to eclipse.

In the aircraft, the telescope was locked on to Uranus well before the expected time of occultation, and all the equipment was checked, found to be working perfectly, and left switched on while the crew relaxed, chatted and waited for the event. Suddenly, the photometer showed a sudden, pronounced drop in light level from the star; seconds later the light level came back up to normal. A sharp-eyed crewman noticed this unexpected dip, and initially put it down to an electronic 'glitch' in the equipment. Minutes later, another dip took place. This time all the crew were watching, and there was no possibility of there being a 'glitch'.

Three more brief but unmistakable dips were seen before Uranus drifted in front of the star: the aircraft was already buzzing with debate about what could have caused the dips. It was immediately obvious that the dips must have been the result of something near Uranus passing in front of the star. That something could only have been a series of previously unknown moons, *or a set of rings*. There was some immediate evidence for that later suggestion, because the drops in light level had not been complete, as would have been the case if a solid satellite had eclipsed the star.

The crucial test was whether or not the same sequence of events took place when the star reappeared on the far side of Uranus. The star was occulted for twenty-five minutes; after it reappeared, the crew noted another five 'secondary' occultations. When they got their feet back on Earth and had time to work out their results, they found that the time intervals between the 'dips' in light level were symmetrical about Uranus. This was clear evidence that Uranus is surrounded by a set of five rings, of which the outermost one is much the strongest. Proof was supplied by the Perth Observatory; there a complete occultation had not been observed, because the city was too far north, but the same series of secondary occultations had been observed. So

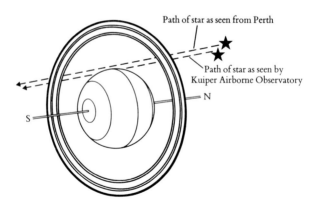

Figure 9.7. Uranus' rings and the apparent path of the star whose occultation led to their discovery. The rings look elliptical in this diagram, since they were turned slightly away from the Earth. Two pairs of thin rings and a more pronounced outer one are shown. Four others are known to exist.

Uranus became the second planet in the solar system known to have rings.

Naturally, the discovery of the rings triggered off a greal deal of follow-up work. The chief conclusions to emerge from these studies were that there are *nine* rings rather than five, and that the material in them has a very low albedo. This is particularly interesting, since it indicates that Uranus' rings *cannot* be made of ice like Saturn's, but must be composed of stony material, a conclusion which is at variance with theoretical predictions. It also seems that the rings are held in their positions by orbital resonances with the satellites of Uranus, which channel particles into tight ring patterns. Further study of Uranus and its rings will undoubtedly help in the development of theories about the formation of planets and their satellites. As the discovery of Jupiter's small ring system emphasized, rings must now be regarded as normal, rather than exceptional, parts of a planetary system.

Uranus has five satellites: Miranda, Ariel, Umbriel, Titania and Oberon (readers may recognize a Shakespearean ring to these names). The largest of these is Titania, with a diameter of about 1,800 kilometres. Because they are so small and so distant, almost nothing is known about these satellites, save that they rotate in the plane of Uranus' equator. Since Uranus' equator is nearly at right-angles to its orbital plane, this means that the satellites corkscrew their way through space in a most unusual manner. It is clear that whatever process was responsible for Uranus lying on its side, the same process affected the satellites. One suggestion is that the planet was 'knocked over' by a major impact with a smaller body, and that the satellites may be fragments of that body.

Neptune

The discovery of Neptune is one of the classics in the history of scientific achievement. It will be recalled that Uranus had been observed, unknowingly, several times before its eventual discovery in 1781. Once its existence had been recognized, astronomers dug up the records of the earlier 'non-sightings' in order to refine the details of Uranus' orbit. They found that they simply could not calculate a workable orbit for the planet: it was always getting ahead of, or behind, the position calculated for it.

In 1834, it was realized that this situation could be accounted for if there were *another* planet, further out than Uranus, whose gravitational effects were perturbing, or distorting, Uranus' orbit. Sometimes the unknown planet would be ahead of Uranus, tending to accelerate it, and sometimes behind it, slowing it down.

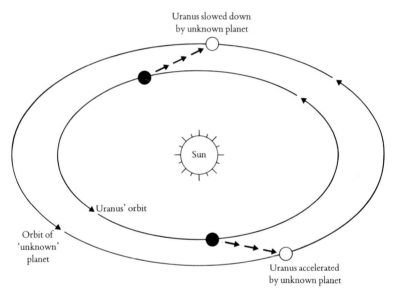

Figure 9.8.   The perturbations of Uranus' orbit caused by an 'unknown' planet.

Once its existence was suspected, a formidable mathematical task faced astronomers. There were simply too many unknowns. The mass, distance and orbital velocity of the 'new' planet all had to be found simultaneously. Apart from the basic mathematical laws of Kepler and Newton, the only guide available was Bode's Law, which gave a rough indication of how far out from the Sun the planet *might* be.

First to tackle this daunting problem was a brilliant Cambridge student, John Couch Adams. He embarked on the task immediately after graduating from Cambridge in 1843, and by 1845 he had worked out where he believed the planet to be. He sent his results to the then Astronomer Royal, Sir George Airy, in the hope that the latter would instigate a telescopic search for the planet. Airy, however, was totally uninterested, did not pursue the matter himself, and discouraged Adams from doing so.

Meanwhile, the equally brilliant young French astronomer, Urbain Le Verrier, had been addressing himself to the same problem, and in 1846 published his calculations, and an estimated position almost identical to Adams', in a memoir to the French Academy of Sciences. Back in England, Airy received news of Le Verrier's work, and, probably feeling more than a little hot under the collar by now, decided that it was high time to do something about Adams' calculations.

He asked James Challis, Professor of Astronomy at Cambridge, to start making a telescopic search for Adams' planet. Sadly, the British astronomer fumbled his chance of finding the new planet. He did not have a reliable chart of the right piece of sky, and therefore started half-heartedly on the tedious job of charting stars from scratch, so that if the new planet were present it would reveal itself when the positions on the star chart were rechecked at a later date.

Le Verrier, of course, was not standing idly by. He had sent details of his work to Johann Encke at the Berlin Observatory, where a search was immediately begun by J. G. Galle, who already had an accurate star chart. Galle received Le Verrier's calculated position on 23 September 1846. The very same evening he found an unknown object less than 1 degree from the calculated position, and two days later there was no room for doubt: Neptune had been discovered.

Although it was Galle who first actually saw the planet, it was Le Verrier's calculations that enabled him to do so. But Adams' work was just as good, and had been completed earlier. So who gets the credit for discovering Neptune? Happily, Adams and Le Verrier are usually equally honoured in scientific texts.

It would be rewarding to be able to follow up the story of the discovery of Neptune with descriptions of some startling new Neptunian phenomena. Sadly, this is not the case. Neptune is so difficult to observe that there is very little indeed that one can say about it, except that it appears to be similar to Uranus.

Neptune's mean distance from the Sun is about 4,500 million kilometres, or 30·1 astronomical units. (Bode's Law predicted that the distance should be 38·8 AU.) Its diameter is 49,500 kilometres, almost identical to Uranus, which means that it is a large planet, and its density is about $1·7 \times 10^3$ kg m^{-3}. This is greater than Uranus', and suggests that it may have more rocky material in its internal structure. There are two known satellites, Triton and Nereid, but there may be other, smaller ones yet to be discovered. The best measurements of the diameter of Triton suggest that it may be nearly 6,000 kilometres across. If these measurements are correct, Triton is the biggest satellite in the solar system, and is 1,000 kilometres bigger than the planet Mercury. (Triton, however, is probably much less dense than Mercury, and hence contains less mass.) Like Titan, Triton is thought to have an atmosphere of methane. It has a *retrograde* orbit around Neptune, an oddity whose relevance will become apparent shortly.

Not surprisingly, no surface details can be seen on Neptune. The only useful observations that can be made are spectroscopic studies of the atmosphere. Like Uranus, Neptune's atmosphere consists mostly of hydrogen and methane, and the methane provides a certain amount of surface warmth because of its greenhouse effect. Temperature studies show that Neptune emits *twice* as much heat as it receives from the Sun. This is quite different from Uranus, and indicates that it must have a source of internal heat. Exactly what process provides this built-in central heating is not known.

One other statistic on Neptune is worth mentioning. It revolves around the Sun in 164·8 years, and it rotates on its axis every sixteen hours. A Neptunian 'year' would therefore be 90,228 Neptunian days long. This would at least mean that a resident of Neptune would have a decent breathing space between income-tax demands . . .

Pluto

The saga of the search for Pluto is in many ways even more stirring than that for Neptune. Almost immediately after the discovery of the latter, Le Verrier himself suggested that there might be yet another planet waiting to be found. In subsequent years, other astronomers began to investigate the problem and to come up with suggestions. By the 1880s, it was being seriously proposed that there were *two* other planets, one at 500 and another at 1,000 astronomical units from the Sun, to account for some remaining unexplained perturbations in Uranus' orbit.

The computations involved were fiendish. The perturbations in Uranus' orbit caused by Neptune were quite large, and once the effect of Neptune was taken into account, the remaining apparent discrepancies were small. Neptune's orbit, of course, would be much ·more perturbed by 'Planet X' than Uranus, but since Neptune had not moved far since its discovery, the details of its orbit were not accurately known. The astronomer who took on the task of finding 'Planet X' has already figured predominantly in this book: he was Percival Lowell, whose work on Mars was so influential. In 1902, Lowell published his belief in the certainty of the existence of a planet beyond Neptune, and in 1905 he committed himself and his observatory to a search for it.

He decided on a 'belt and braces' approach. He started both to look for the planet directly, making photographic searches of the sky, and also to try to calculate its position from theoretical considerations. For more than ten years he persevered. He drove himself unceasingly, and relentlessly chivvied his staff at the Lowell Observatory at Flagstaff. At one time, he had four people employed full time as a kind of human computer to deal with the endless series of calculations. While away from the observatory, he bombarded the staff with fresh suggestions of where to look, and demands for further photography in the light of his revised calculations. All was in vain: 'Planet X' remained undiscovered.

In 1915, Lowell published a memoir summarizing his ten years' work, presenting a careful appraisal of the 'Planet X' problem, and giving two alternative possible positions for the planet. The following year he died, his life's ambition frustrated. According to his brother, Lowell's failure to discover 'Planet X' was 'the sharpest disappointment of his life'.

Fourteen years later, on 13 March 1930, Lowell's successor as director at his observatory, V. M. Slipher, was able to announce the discovery of the new planet. The search had been suspended for many years after Lowell's death. When it was resumed in 1929, it took only a few months for the discovery to be made, using a new 13-inch telescope which had been constructed especially for the purpose. A young assistant at the observatory, Clyde Tombaugh, was assigned the task of systematically photographing regions of the sky along the ecliptic at two- or three-day intervals, and then comparing pairs of photographs of the same region with an instrument known as a link comparator, which revealed immediately any object which had shifted its position between the times of the two exposures. After many tedious hours at the comparator, Tombaugh found what he was looking for: a tiny speck of light which had moved the correct distance between two exposures.

After tracking the object to establish its identity beyond doubt, the observatory made the following announcement by telegram to astronomers all round the world: 'Systematic search begun years ago supplementing Lowell's investigations for Trans Neptunian planet has revealed object which since seven weeks has in rate of motion and path consistently conformed to Trans Neptunian body at approximate distance he assigned . . .'

Surprisingly, in view of the complexity of the calculations involved, the planet was found only 6 degrees away from the position Lowell had predicted for it, a great vindication of his years of patient effort. It was a tragedy that he was denied the experience that Keats had in mind when he wrote his famous lines from 'On First Looking Into Chapman's Homer':

> Then felt I like some watcher of the skies
> When a new planet swims into his ken . . .

Fittingly enough, the planet was christened *Pluto*, after the god of the Underworld. Located on the frigid outer fringes of the solar system, the planet must indeed be an unimaginably desolate, hostile place and a most appropriate home for Pluto. More to the point, the name was chosen so that its first letters, PL, commemorate *P*ercival *L*owell.

Once the excitement of the discovery had died down and astronomers had a chance to go through old records, it was found that Pluto had been recorded on no less than *sixteen* previous occasions. Ironically, on two of these occasions, the planet had been photographed at the Lowell Observatory during Lowell's lifetime, but on both occasions had been misidentified as a star.

It is nearly half a century since Pluto was discovered, but not much more is known about it now than when it was first found. Its mean distance from the Sun is 5·9 billion kilometres, but its orbit is so elliptical that it sometimes passes within the orbit of Neptune. Its orbital plane is inclined at an angle of 17 degrees to the ecliptic, twice the inclination of any other planet. Pluto and Neptune provide yet another example of orbital resonance: Neptune orbits the Sun exactly three times in the time that Pluto takes to go round twice.

Figure 9.9.   Two photographs of Pluto taken twenty-four hours apart, showing the motion of the planet. Although taken with the massive 200-inch Mount Palomar telescope, the planet appears as a star-like point source of light. (Hale Observatory photograph)

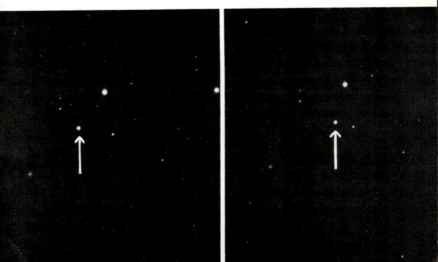

Since it scarcely shows a disc in even the biggest telescopes, the diameter of Pluto is exceedingly difficult to measure. The *mass* is even more difficult to determine, and is more critical, since this is the key to establishing what the planet is made of. Some early work suggested that it should have a mass seven times that of the Earth, in order to account for perturbations in Uranus' and Neptune's orbits. Recent work discounted any major perturbations, and suggested that Pluto is much smaller than the Earth.

The first new data on Pluto for many years emerged in 1978, as a result not of spacecraft investigations but of straightforward telescopic work. J. W. Christy of the US Naval Observatory was studying photographs of Pluto taken with the observatory's 155-centimetre telescope. He noticed that Pluto's disc was not circular, but appeared to be elongated. Checks with previous pictures confirmed this elongation, so that the possibility of photographic distortion could be eliminated, and further confirmation was provided by the Cerro Tololo telescope in Chile. The strange elongation of Pluto's image was interpreted as being caused by a large moon orbiting around Pluto, so close to the planet that the two bodies cannot be separately resolved.

The new moon has been (unofficially) christened *Charon*, a most appropriate name, since Charon was the mythological boatman who ferried the souls of the dead across the river Styx to the Underworld. The best estimates suggest that Charon has a diameter 40 per cent of Pluto's (making it much the largest satellite relative to its parent body in the solar system), and orbits at a distance of only 17,000 kilometres from Pluto.

The beauty of the discovery of Pluto's moon is that it provided much sorely needed data on the planet itself. In particular, the mass could be determined for the first time, and was found to be 1.5×10^{22} kg. The diameter is about 2,700 kilometres and the density about 1.5×10^3 kg m^{-3}. Pluto is therefore a tiny planet, the smallest in the Sun's family, and is probably made of ice and frozen volatiles. Observation of Charon's period of revolution around Pluto revealed one final, uniquely elegant example of spin-orbit coupling: the revolution period is the same as Pluto's rotation period (6·4 days). Charon is thus locked permanently above the same point on Pluto's surface, a moon that never sets or rises.

Because Pluto is such a small scrap of a body, hardly worthy of the status of a planet, it has been argued that it is not a planet at all, and may be an escaped satellite of Neptune. One possible scenario suggests that Pluto and Triton once revolved around Neptune in conventional orbits, until Triton's orbit brought it so close to Pluto that it whipped around it and went back in the opposite direction, and in so doing caused Pluto to be accelerated and flung off into the depths of space, where it eventually settled into orbit around the Sun. This explanation has the advantage of simultaneously explaining Pluto's orbital peculiarities, as well as Triton's retrograde revolution. It does not explain how Charon appeared on the scene, unless the catastrophic event which flung Pluto out of Neptune's embrace also caused Pluto to break up into two smaller pieces.

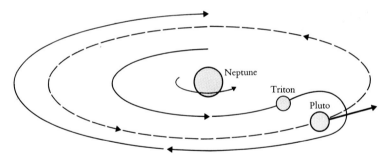

Figure 9.10. Hoyle's explanation of how interaction between Triton and Pluto might have caused Pluto to leave Neptune altogether, and Triton to enter a retrograde orbit around Neptune.

Pluto is always thought of as the most distant member of the solar system. In January 1979, however, its orbit brought it swinging in nearer the Sun than Neptune, and for the following twenty years Neptune will have the distinction of being the most remote planet. Notwithstanding this minor quirk, with Pluto we arrive at the dim, frigid limits of our knowledge of the planets.

There may be other bodies further out than Pluto, silent slaves to a Sun so distant that from them it would appear only as a bright point of light in a dark, starry sky. If such bodies do exist, they are probably small, insubstantial chunks of ice doomed to remain tumbling slowly in their orbits until, decades or centuries from now, a drifting spacecraft, aged by the long years of its flight, carries out a silent electronic survey and transmits images of them to our descendants on Earth.

10. Meteorites, Asteroids and Comets

Meteorites

Most of us will never lay eyes on Uranus, Neptune or Pluto, but it is not difficult to experience the pleasure of observing bits of debris from distant parts of the solar system burning up in the Earth's atmosphere. An hour spent on any clear, dark night will probably be rewarded with the sight of at least one 'shooting star', a quick slash of light across the sky that is over almost before its presence has been registered. Systematic observation over a period of months may be rewarded with a glimpse of a much more spectacular 'fireball' blazing brilliantly as it sweeps silently far overhead.

Shooting stars – or *meteors* – have been known since antiquity. (*Meteorology*, by the way, has nothing to do with meteors, except that both words are derived from the same Greek root.) Accounts of stones mysteriously 'falling from the sky' can also be found in many of the earliest historical records. Amazingly, though, it was not until the end of the eighteenth century that it was realized that there might be a link between 'shooting stars' and the mysterious arrival of stones from nowhere, or that these stones might be derived from outside the Earth. This discovery was first reported authoritatively by a Russian scientist, E. F. Chladni, in 1794. A few years earlier, three French Academicians, including the chemist Lavoisier, had dismissed a stone actually *seen* to fall at Luce in 1768 as merely an ordinary stone that had been struck by lightning.

In 1803, the sceptics were finally forced to reconsider their views. On 26 August of that year there was a violent explosion in the neighbourhood of the village of L'Aigle in France, which was audible at a distance of over 120 kilometres; a brilliant fireball was seen simultaneously, and the neighbourhood of L'Aigle was showered with no less than *3,000* meteorite fragments. The L'Aigle event was studied by the French physicist, I. B. Biot, and it could be said that this great shower forced the scientific study of meteorites into existence.

Naturally, it took many years for the cosmic origin of meteors and meteorites (the objects that reach the ground) to be universally accepted. In 1808, Thomas Jefferson, third President of the United States, was still clearly unconvinced. Writing of an alleged fall of stones from the sky, he declared: 'It may be very difficult to explain how the stone . . . came into the position in which it was found. But is it easier to explain how it got into the clouds from whence it is supposed to have fallen?'

Meteoritics is now an important branch of science in its own right. Most meteorites are not particularly impressive to look at, since they are either drab, brownish stones indistinguishable to the untrained eye from any other stone, or rusty, irregular lumps of iron that would easily pass for scrap metal. The unique importance of meteorites, which makes planetary scientists prize them highly, is that they are tangible samples of material of extra-terrestrial origin. Before the Apollo missions, meteorites provided the only *direct* information on what the components of the solar system might be made of. In the wake of the Apollo missions, meteorite studies are continued more enthusiastically than ever, because in them is preserved evidence of some of the earliest events in the origin of the planets, evidence which even the Moon is too young to retain.

The 'shooting stars' that are so easy to observe on any dark night are caused by tiny particles of inter-planetary dust entering the Earth's atmosphere and burning up. Such events are taking place all over the Earth, all the time. Many experiments have been carried out to find out just how many such *micrometeorites* strike the Earth each day, but it is not easy to get good results. One interesting statistic suggests that one particle weighing one millionth of a gram will impact on each square metre of the Earth's surface once every hundred days. Another is that the rate of meteoritic infall on to the Earth is at least 100 tonnes per day!

A large quantity of this dust-like material reaches the ground unnoticed. It accumulates in areas where geological processes are slow, as in deep-sea sediments and the polar ice-caps, and is increasingly coming under study as more sophisticated analytical techniques become available. Much more is known about the larger meteorites, which are big enough to have an individual identity. The statistics again vary, but there are probably something like 500 such arrivals a year over the whole Earth. About 350 of these disappear into the oceans, to be lost forever, and only 150 hit dry land. Of these, less than ten per annum are found and passed on to scientists for study.

The small number of specimens whose arrival from space is actually observed to happen are known as 'falls'; they are particularly valuable. Rather more common are the so-called 'finds': meteorites which fell at some unknown date in the past and which have come to light accidentally, sometimes being turned up by a plough in a field, or more rarely as the result of a deliberate search. In recent years, the Antarctic has provided a treasure trove for meteorite hunters. Here, meteorites falling over the centuries into the ice surface have been carried slowly forward to areas on the glaciers where melting and ablation of the ice take place. This provides a highly efficient natural process for concentrating meteorites at a single point, and many important finds have been made by relatively cursory examinations of some obvious 'concentration points'.

Some recent falls have been uncomfortably large. About 2,000 kilograms have been recovered from one meteorite that came to ground near Pueblito de Allende in Mexico in 1969, and a similar weight fell near the city of Jiling, China, in March 1976. If such a large body were to fall *on*, rather than near, a city, it could cause a great deal of damage. There is, however, surprisingly little evidence of anyone ever having been hurt by a meteorite fall. Apart from a few dubious accounts of people being killed in the distant past, there is only one record of anyone being hit: in 1954, a housewife in Alabama was struck on the arm by a meteorite which crashed through the roof of her house and bounced off a radio. She was more surprised than hurt!

The largest 'finds' are much larger than the 'falls'. The record is held by the Hoba meteorite, found on a farm in South-West Africa (now Namibia). This is so heavy that it has not yet been moved from the place where it fell. It consists of a solid mass of iron weighing at least 60 tonnes, and may have weighed over 100 tonnes when it first fell. Many other large iron meteorites weighing several tonnes are known around the world. Some were probably discovered in antiquity, because there is evidence of them being used as a source of iron for tool-making by civilizations that had not learned to make iron themselves. Groups of Greenland Eskimos were found to be using tools of meteoritic iron when contacted by the early explorers.

When they enter the Earth's atmosphere, meteorites are travelling extremely fast, with velocities of the order of kilometres per second. As they plunge deeper and deeper into the atmosphere, the smaller meteorites are slowed by atmospheric drag and are simultaneously heated up by the friction of the rush of air past them. The temperature may rise by thousands of degrees, producing glowing fireballs and causing many of the smallest particles to burn up entirely.

As recently as 1972, a large fireball was observed by people on the ground and tracked by radar passing over the state of Montana in the United States. The meteorite may have weighed as much as 1,000 tonnes. Fortunately, it seems to have skipped off the Earth's atmosphere, like a 'ducks and drakes' pebble off a pond, and did not get closer than 60 kilometres to the ground. Had it approached at a slightly steeper angle, it might have crashed down in Alberta, Canada, blasting out a 100-metre crater. It does not require much imagination to appreciate what would have happened had the meteorite come down in a populated area . . .

As it is, the atmospheric braking effect is such that even lumps weighing a kilogram or so are travelling rather slowly when they hit the ground, so that they rarely do much more than bury themselves a few centimetres deep, or, as at Barwell in England on Christmas Eve 1965, break through roofs and floorboards. As the size of the body increases, however, so the braking effect decreases dramatically, and really large bodies – weighing many tons – are hardly slowed up at all. The result is that they impact the surface at near-cosmic velocities and explosively excavate craters.

Figure 10.1. The impact crater left by the Barwell meteorite when it made its uninvited arrival on the driveway of a suburban house on Christmas Eve 1965. (Institute of Geological Sciences photograph)

The best known of these is Meteor Crater in Arizona, a strange, bowl-shaped depression blasted out of the flat Arizona plains some 20,000 years ago by an iron meteorite believed to have weighed about 30,000 tonnes. Rusting bits of meteorite debris can still be found around the rim of the crater, and large lumps of it have been recovered from the surrounding area, but no single, large mass has been identified, despite intensive research. It seems that most of the mass of the meteorite was simply vaporized during the instant of impact.

Figure 10.2.   Undoubtedly the most spectacular impact crater on Earth, Meteor Crater in Arizona is about 1·2 kilometres in diameter and 100 metres deep. Its shape is not perfectly circular, but more like a rounded square. This is a result of fracture patterns in the target rock.

Figure 10.3.   Small shaly fragments of oxidized iron meteorite collected from the rim of Meteor Crater. The largest is 5 centimetres across.

The number of impact craters in the world as a whole is surprisingly small – less than a hundred are well authenticated. By contrast with the Moon, Mars and Mercury, where impact craters are by far the most important surface features, terrestrial impact craters are so rare that they rate as curiosities. This is not only because of the shielding effects of the Earth's atmosphere, which burns up all but the largest bodies, but is also caused by the constant erasure of surface features by geological erosion. As discussed earlier, the main episodes of impact events on planets took place very early on in the history of the solar system, more than 4 billion years ago. The Moon retains a full record of that frightful bombardment; on Earth, the massive craters that were formed then have long since been smoothed away.

It is easy enough to conclude that the Moon's surface is older than the Earth's, simply because it has more impact craters, but making this kind of comparison rests heavily on the assumption that the Moon has had the same history of bombardment as the Earth. It is generally assumed that the rate of bombardment for each has drastically decreased compared with the initial rate, but it is impossible to be certain that they have always shared the *same* rate of bombardment. Even greater problems arise when trying to draw comparisons with much more distant bodies, such as Mars. The subject is an important one, because it is through the process of bombardment that the planets originally came together, or accreted, from the cloud of dust particles that made up the primordial solar system. Some of those particles could be identical with the micrometeorites still arriving on Earth.

Because they are so important in understanding the formation of the solar system, it is not surprising that the chemical composition of meteorites has been exhaustively studied. One of the delightful things about meteoritics is the strictly defined data base that the whole subject is built up on. In 1966, for example, there were exactly 1,791 known meteorites, of which 56 per cent were 'finds' and 44 per cent 'falls'. Many have been added since then, of course, and many others probably lie forgotten and uncatalogued in the basement storerooms of museums around the

Figure 10.4. Manicougan Crater, Quebec, Canada. This 65-kilometre diameter crater is so large that it can only be clearly seen on satellite photographs. The crater is believed to have been formed by an impact taking place 214 million years ago. (LANDSAT photograph)

world, but the total is still probably below 3,000. (In these figures, fragments falling in a shower are counted as a single meteorite.) Meteoriticists, therefore, are always eager to get their hands on new specimens: perhaps they live in the hope that the next meteorite to fall out of the blue will be of a kind new to science. It is by no means a wildly unrealistic hope – the Allende meteorite of 1969 yielded some major new clues to the origin of the solar system.

There are many different kinds of meteorite, and the terminology used to describe them is complex. For our purposes we can think of them in terms of three main, self-descriptive types: irons, stony irons and stones.

Irons are much the best-known meteorites, and make up more than half of the known 'finds'. This is scarcely surprising, because a lump of iron in the middle of a field is bound to attract more attention than a nondescript rock. Irons, however, form only about 5 per cent of all falls; they therefore represent only a small proportion of the population of meteorites whirling round the solar system.

Irons are not pure iron. They usually contain several per cent of nickel, and traces of cobalt, sulphur, phosphorus, copper, chromium and carbon. When cut, polished and etched, iron meteorites reveal a beautiful internal texture of intergrown crystals of metallic iron. This is called the *Widmanstätten texture*, after its Austrian discoverer, Count Alois de Widmanstätten. Apart from its aesthetic attraction, this texture is valuable because it is unique to meteorites and enables one to distinguish instantly between an iron meteorite and a random lump of scrap metal.

Iron meteorites have already played a major part in science: for many years, one of the chief lines of evidence that the Earth has a metallic core has been that iron meteorites have almost

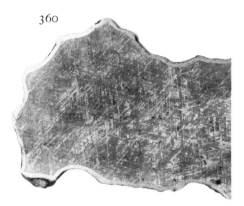

Figure 10.5. An etched and polished slice of an iron meteorite showing the beautifully regular Widmanstätten texture, which is diagnostic of meteorites. Such textures would take millions of years to produce in terrestrial iron melts. (Institute of Geological Sciences photograph)

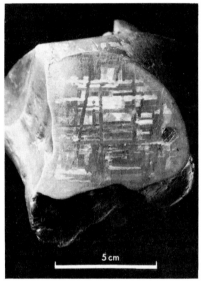

5 cm

Figure 10.6. The Rowton meteorite, a beautiful iron which fell in Shropshire in 1876 and is the only known 'fall' of an iron in the British Isles. Widmanstätten texture is visible on the polished face. (Photo courtesy of the trustees of the British Museum, Natural History)

exactly the right properties, and their existence indicates that material which could have contributed to the formation of the core is still around in space.

Stony irons are, as their name implies, mixtures of iron and stony material. The iron is the same kind of iron–nickel alloy as in the irons, while the stony material consists of silicate minerals such as olivine and pyroxene, well known in terrestrial rocks. They are rather rare, forming only 2 per cent of all falls.

Stony meteorites are much the most common type, forming 93 per cent of all falls. Although not so interesting to look at as the irons, there is more to them than at first appears. They have a distinctive texture, made up of small spheres of silicate minerals embedded in a fine-grained matrix. These spheres are known as *chondrules*, and meteorites possessing them are called *chondrites*. Chondrules are thought to be quick-frozen drops of liquid silicate melt, formed at the time of formation of the meteorites themselves. How they came to be molten in the first place is not understood.

Figure 10.7. One of the fragments of the Barwell meteorite. The stony composition is well seen on the fractured surface. The outer surface shows the smoothing and polishing caused by frictional heating and ablation during the meteorite's descent through the atmosphere. (Institute of Geological Sciences photograph)

There are many different kinds of chondrite meteorite, most of them having compositions roughly like that of peridotite, of which the Earth's mantle is made. Much the most important type, however, are the so-called *carbonaceous chondrites*. Although exceedingly rare, these strange objects have provided scientists with a wealth of controversial data. The meteorites contain a few per cent of carbon, which is not so unusual, but they also contain traces of organic carbon compounds used in life-forming

Figure 10.8.   The radiating linear features on the surface of this meteorite were formed during its descent through the atmosphere. Known as a flight-orientated chondrite, the meteorite adopted a stable altitude in its descent, a liquid fusion crust built up on its leading edge as the result of frictional heating, and the slip-stream smeared the crust out radially. The 12-centimetre diameter meteorite fell at Ashdon, Essex, in 1923. (Photo courtesy of the trustees of the British Museum, Natural History)

processes. A good example is the Murchison meteorite, which thundered into the scientific consciousness on 28 September 1969, over the town of Murchison, Australia. This was only the twentieth known specimen of a carbonaceous chondrite, and thus a great scientific rarity; but it was not very impressive to behold, collectors who recovered the fragments describing it as a crumbly, carbon-rich clay. The chief interest in the meteorite lay in its chemistry. Analysis revealed that it contained traces of seventeen different fatty acids and eighteen amino acids. These, when woven together properly, constitute the foundations of cellular life.

Similar chemical compounds have been discovered in the Orgueil meteorite, and it has even been claimed that 'fossils'

Figure 10.9.   A fragment of the Murchison meteorite about 10 centimetres across. The fractured surface shows light-coloured chondrules and some important Ca-, Al-, Ti-rich inclusions in a dark carbonaceous matrix. (Photo courtesy of the trustees of the British Museum, Natural History)

resembling algae have been found in it. Such 'discoveries' have led to claims that life on Earth may have been 'seeded' by 'spores' arriving on meteorites.

This is not a widely accepted view. There are many ways in which quite complex organic compounds can be made by entirely non-biological processes. Much more important, and much less controversial, is the suggestion that carbonaceous chondrite meteorites are extremely *primitive* objects; they are as near as we can get to samples of the material from which the solar system evolved. This suggestion is based on the fact that the composition of the carbonaceous chondrites is closely similar to that of the Sun, although the light gases such as hydrogen and helium are not present. Most of our discussion of the origin of the solar system starts with the assumption that the primordial material, from which everything else was derived, had a composition roughly that of carbonaceous chondrites.

It might be argued that just because a meteorite has the same composition as the Sun does not mean that it has anything whatever to do with the origin of the solar system: the meteorite might have been formed yesterday. But the other unique characteristic which makes meteorites vitally important in discussions of the origins of the solar system is that they are *all* extremely old. Before samples were obtained from the Moon, meteorites were the oldest known objects that Man had laid his hands on. Sophisticated radiometric dating techniques have shown that meteorites formed about 4·6 billion years ago, and this is generally accepted as the age of the solar system as a whole.

It might seem, then, that all meteorites are extremely primitive, that they are just odd bits of junk left lying around in space after the planets formed. This is not the whole truth. Careful physical and chemical studies of meteorites show that most of the stones and irons could not have formed originally in their present small sizes; they must be fragments of much larger bodies, perhaps between 200 and 500 kilometres in diameter. Furthermore, studies of the effects on meteorites of cosmic rays (similar to the 'sun tan' studies of lunar samples) have revealed that many have *not* existed as individual entities since their formation, but seem to have become separated from a parent body some millions of years before their arrival on Earth. Since they have the right dimensions, it is natural to think of the asteroids as the parent bodies. So is there any connection between meteorites and asteroids? Where in space do meteorites come from? They cannot simply arrive out of nowhere!

Finding out where meteorites come from has been, and is, a grave problem for astronomers, because they arrive so quickly. Ideally, what is needed is photographic coverage of the fall of *one* meteorite from *two* widely separated observation sights so that an accurate trajectory can be calculated. This has been successfully done for two meteorites, one of which fell near Pribram, Czechoslovakia, in 1959, and one at Lost City, Oklahoma, in 1970. Re-

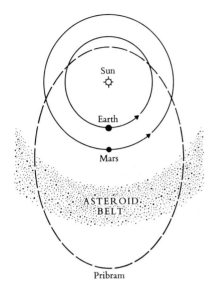

Figure 10.10.   The orbit of the
Pribram meteorite, determined by
photographic studies of its fall.
Since these relate to only the last
few seconds of the meteorite's flight,
there are clearly possibilities
of large errors in the trajectory.

coveries of material were made in both cases, and the calculated trajectories showed that the meteorites hit the Earth because they had been travelling in elliptical, Earth-intersecting orbits around the Sun. As Figure 10.10 shows, the sweep of the orbits of these two meteorites carried them far away from the Sun, out towards Jupiter. This is exactly the domain of the asteroids. Which leads us nicely on to a discussion of the asteroids themselves.

The asteroids

The role of asteroids in the history of the solar system studies has been briefly mentioned in Chapter 1, where the importance of the asteroids in filling the 'Bode's Law Gap' between Mars and Jupiter was discussed, and the usefulness of one of the asteroids (Eros) as a stepping stone in calculating the numerical value of the astronomical unit described.

The first asteroid to be discovered, Ceres, was found on 1 January 1801 by the Sicilian astronomer, Giuseppe Piazzi. At the time, a coordinated search was being carried out by European astronomers for the so-called 'missing planet', but it seems that Piazzi found Ceres more or less by accident, while he was mapping the stars in the constellation Taurus. He noticed a point of light that did not seem to belong to Taurus, and subsequently found that the star-like object was not stationary but moved against the starry background.

Ceres is now known to be much the largest of the asteroids, with a diameter of about 955 kilometres. Since Piazzi's day, smaller asteroids have been discovered at an ever-increasing rate. It is now thought that there are at least *70,000* of these objects, most of them less than 100 kilometres in diameter. For a long while, asteroids were not regarded with much enthusiasm by astronomers, because they are so small and so abundant. Walter Baade of the Mount Wilson and Palomar Observatories even dismissed them scornfully as 'vermin of the sky'. Others, how-ever, perhaps driven by the same motives that compel some individuals to collect matchboxes, beer mats or spiders, went out of their way to find and 'collect' new asteroids. One of the perks of this occupation was the opportunity it gave the discoverers to give their heavenly body the name of their choice. So, while many asteroids have the usual names from classic mythology – Pallas, Juno and so on – some have much more down-to-earth names, including those of the discoverers' wives or current girl friends.

More recently, asteroids have received more serious attention, and are likely to continue to do so in the future, particularly because they do offer realistic targets for spacecraft investigations – a spacecraft could land and take off again from an asteroid much more easily than from a full-sized planet.

One of the most aesthetically pleasing facets of asteroids is their distribution within the solar system. Not only do they constitute the well-known 'asteroid belt', plugging the gap between Mars and Jupiter, but they also show some much more subtle features. Careful work over the years has shown that there is not simply a single great cloud of asteroids in the belt, but that there are marked concentrations at particular distances from the Sun. Figure 10.11

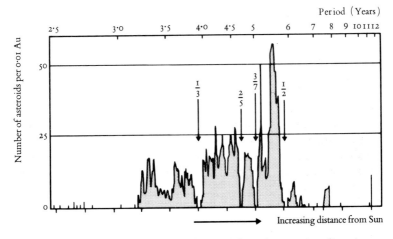

Figure 10.11.   The distribution of the asteroids with increasing distance from the Sun. The orbital periods are shown along the top, and the gaps where resonances exist between the orbital periods of the asteroids and Jupiter are arrowed.

summarizes the distribution. At first sight this diagram looks thoroughly erratic. Closer inspection shows that there are a number of clear 'gaps' where no asteroids occur, and that these gaps are rather elegantly arranged.

The asteroids, like all other objects in the solar system, obey Kepler's Laws. Thus, the further away from the Sun they are, the longer their period of revolution about it. The gaps occur at distances from the Sun *where the asteroids' periods would be in a simple ratio to Jupiter's period.* Thus, there are no asteroids at the points where the ratio between periods is 3:1, 5:2 or 2:1. This harmonious relationship is brought about by resonances between the gravitational field of Jupiter and that of the Sun; the 3:1 resonance marks the point at which an asteroid would go round the Sun three times in the time it takes Jupiter to go round once. The gaps are called *Kirkwood gaps*, after their discoverer.

Another elegant relationship was predicted in 1772 by the French mathematician, Joseph Lagrange. He calculated that the gravitational interactions between the Sun and Jupiter were such that there should be two points on Jupiter's orbit, 60 degrees ahead and 60 degrees behind the planet, where small bodies could orbit in dynamic stability. These points are known as *Lagrangian points*. In 1905, over a century after Lagrange's prediction, the first asteroid was discovered occupying one of the Lagrangian points. It is now known that there are clusters of asteroids at each point, and that, for some reason (not explained by Lagrange), there are twice as many occupying the point ahead of Jupiter as there are the one behind. The asteroids are known collectively as the Trojans, and the two stable points as the Greek and Trojan Camps.

There is more to asteroids than mere mathematical niceties, however. What could be more calculated to bring one's thoughts back down to Earth than the brutal fact that some asteroids are moving on orbits which could bring them into collision with the Earth? Before the reader starts worrying about whether his insurance policies regard collision with an asteroid as an 'act of God', or whether, indeed, any insurance companies will survive such a collision, let him be reassured that the risk is not great. Statistically, the chances are much greater that the reader will die of boredom while reading this page than that he will live to experience a collision between the Earth and an asteroid!

The asteroids which present this remote threat to the Earth are not those of the main asteroid belt: their orbits keep them permanently safely out beyond Mars. There is a much smaller group, however, which has highly elliptical orbits, which mean that they swing in from the asteroid belt towards the centre of the solar system, crossing the orbits of Mars, the Earth and Venus and, in some cases, even that of Mercury. The group as a whole are known as the *Apollo asteroids*, after the first individual to be discovered which crossed the path of Venus. Eros is one of the Apollo group, and so is Icarus, whose orbit is so elliptical that it sweeps nearer to the Sun than any other object in the solar system. (Icarus was the figure in mythology who built himself wings with wax and feathers, and then was so carried away with the joy of flight that he flew too close to the Sun; the wax melted, and he

Figure 10.12. The orbits of some of the unusual asteroids that do not fall within the main belt. Only individual examples of asteroid groups are shown. Comparison of this diagram with Figure 10.10 (page 365) shows that the orbit of the Pribram meteorite resembles those of some asteroids.

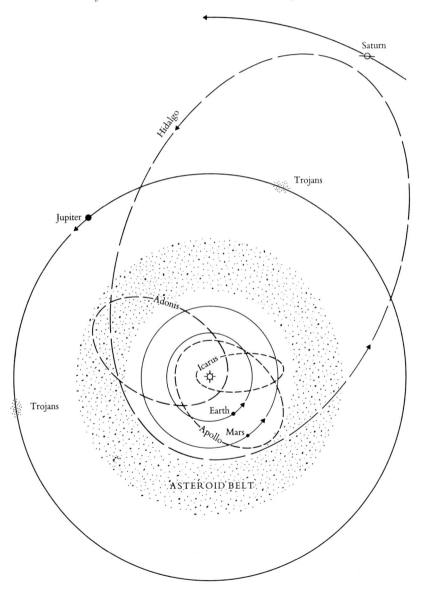

plummeted to his death. *Icarus* is now the name chosen for one of the most forward-looking journals on planetary science.)

In 1937, evidence of how close the Apollo asteroids come to Earth was convincingly given by the small body Hermes, which swept past the Earth at a distance of only 640,000 kilometres. This was a fairly safe miss, of course, but it has been calculated that, if the data on Hermes' orbit are correct – and this is not certain – then it would be possible for Hermes to pass between the Earth and the Moon from time to time. That would be getting uncomfortably close!

About forty asteroids which have Earth-crossing orbits have been observed to date. Estimates suggest that there may be about 1,500 others larger than 1 kilometre in diameter still to be observed. Because they are so small, and come so close to the Earth, it would be theoretically easier, and would require less fuel, for a spacecraft to visit one of these asteroids than the Moon. This offers some immensely exciting prospects, not only because of the opportunity to investigate directly a different member of the solar system, but also because of their potential as open-cast orbiting mines.

The chief difficulty involved in constructing the space stations beloved of science-fiction writers is the enormous quantity of energy required to get the components of the space station from Earth into space. Compare the size of the tiny Apollo capsules to the enormous Saturn rockets required to lift them off the Earth! It has long been argued that this energy cost could be reduced by obtaining the materials for a space station from the Moon, because of its much smaller mass and surface gravity. Brian O'Leary, an astronomer who trained as an Apollo astronaut but who dropped out because he did not like flying, has pointed out that the Apollo asteroids would be an even better bet, since their gravity is almost zero. Because their masses are so small, it might even be possible to hoick such an asteroid out of its proper orbit and 'park' it in Earth orbit, to be used at leisure.

These intriguing speculations all depend on the asteroid being made of something *useful*. So, how does one find out what an asteroid is made of, given that it can only be seen as a tiny point of light in a telescope? This is where the science of remote sensing really comes into its own. Two techniques have been used with great success: *spectrophotometry*, measuring the amount of sunlight reflected by the asteroid at different wavelengths; and *polarimetry*, measuring the extent to which the reflected light is polarized at different angles. Both of these techniques, it will be recalled, were also applied to the Moon in the pre-Apollo era.

Telescopic observations of both the Apollo and main-belt asteroids showed that two main types could be distinguished: one type that is predominantly reddish, and another that is more neutral coloured. Comparison with spectrophotometric data on meteorite samples showed that there are convincing similarities between meteorites and asteroids. The reddish asteroids seem to resemble chondritic and stony iron meteorites, while the others are darker and resemble the carbonaceous chondrites. Within the main belt, there is a marked tendency for the asteroids nearest the Sun to be of the stony iron type, and for the furthest out ones to be carbonaceous chondrites.

The similarity between asteroids and meteorites is no coincidence: it is widely thought that meteorites are derived from asteroids, particularly those in the Apollo group, whose orbits bring them so close to Earth. Meteorites may, in fact, be nothing more than small fragments of asteroids which were broken up by some ancient collision in space. The same collision, or collisions, may also have caused the Apollo asteroids to have been knocked out of conventional main-belt orbits into their present highly elliptical ones.

Several other kinds of information have been gleaned by meticulous studies of asteroids. The masses and densities of three of the largest (Ceres, Pallas and Vesta) have been found. Vesta has a density of about 3.5×10^3 kg m^{-3}, close to that of stony meteorites. Ceres and Pallas are lower: around 2.5×10^3 kg m^{-3}, which is close to that of the carbonaceous chondrites.

It has also been possible to deduce the *shape* of a few asteroids. It is easy to show that many are not simple spheres. If you spin a ball, it looks like a uniform blur. If you spin a coin, you will see a rapid succession of flashes, as the flat sides catch the light. Clearly, more light reaches your eye from the flat sides than from the narrow edge. The same applies to non-spherical asteroids. When their light is measured, the intensity varies consistently, indicating that they are both rotating and non-spherical in shape. In general, the larger the asteroid, the more nearly spherical it is likely to be.

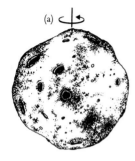

Figure 10.13. (a)　A nearly spherical asteroid will not change in apparent brightness on rotation, unless it has dark and light surfaces.
(b) and (c)　An irregularly shaped asteroid will look brighter when seen broadside on than when seen end on.

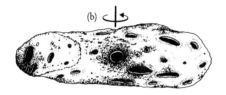

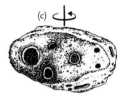

When Eros passed in front of a star, on 23 January 1975, a different kind of opportunity arose to determine its shape. By measuring carefully the shadow zone of Eros with respect to the star, Brian O'Leary and his colleagues were able to show that Eros is far from spherical; it is an irregular slab about $7 \times 19 \times 30$ km.

The occultation of a star by an asteroid led to another important discovery in June 1978. The asteroid in question is known as 532 Herculina, and is a body about 220 kilometres in diameter. Its occultation of the star had been predicted in advance, and careful arrangements were made to observe it. Detailed records were kept by two observatories, and both detected a secondary event taking place *before* the main occultation – in plain language, *another* object had passed in front of the star. This indicated conclusively that Herculina has at least one companion orbiting around it, and possibly more. Herculina's satellite appears to be about 50 kilometres in diameter and is about 1,000 kilometres distant from it.

The discovery triggered off a search for other asteroids with satellites. At the time of writing, at least eight candidates had been located, and it is beginning to appear that asteroids may not be solitary large chunks circling round the asteroid belt but rather clusters of many small bodies. This impression is supported by other studies of asteroids, which suggest that collisions between asteroids in the crowded traffic lanes of the asteroid belt are rather common, and thus that an asteroid probably does not consist of a solid mass of rock, but rather a coagulation of shattered fragments jumbled loosely together in an irregular heap. The term *megarego-lith* has been coined to describe this texture. Clearly, if the asteroids have been effectively reduced to rubble by collisions, it is not too surprising to find that some of the larger fragments knocked off are still orbiting around their parent.

While it may be decades before O'Leary's schemes come to fruition, and spacecraft actually visit an asteroid, it is already possible to suggest what an asteroid may look like. The Viking pictures of Phobos and Deimos, Mars' diminutive satellites, are very instructive. Although these may not be asteroids *per se*, their dark colour suggests that they have the same composition as carbonaceous chondrites, and their heavily cratered surfaces are exactly what one would expect to find on an object that has been battered around for aeons in the asteroid belt.

Comets

> When beggars die, there are no comets seen:
> The heavens themselves blaze forth the death of princes.

These famous lines from Shakespeare's *Julius Caesar* encapsulate many of the early views of comets, the most ostentatious but least known members of the solar system. A major comet is a magnificent sight, one that can hardly fail to make an impression on even the most jaundiced twentieth-century observer. It is scarcely surprising, then, that comets have been written into the histories of many great events. It seems that a great comet was seen at the time of the death of Julius Caesar in 43 B.C.; it is possible that it was a comet that the Three Kings saw when they followed a 'star'

Figure 10.14. Two views of the superb spectacle of Halley's Comet in 1910. The upper picture was taken in Honolulu on 12 May, and shows the comet tail streaming across 30° of the night sky; the lower was taken on 15 May when the tail extended over no less than 40°. (Hale Observatory photograph)

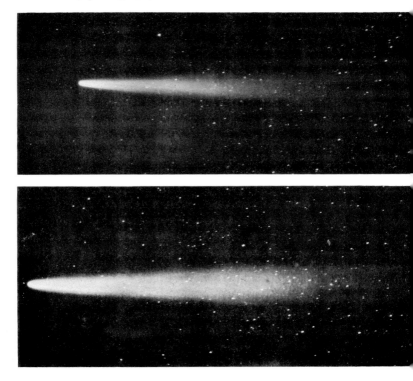

to the birthplace of Christ; it is certain that Halley's Comet blazed forth from the skies during William the Conqueror's uninvited visit to Britain in 1066 – it is depicted in full colour on the near-contemporary Bayeux tapestry.

Halley's Comet is named after Edmund Halley, a contemporary and friend of Sir Isaac Newton. His chief contribution to science was the discovery that comets are bodies which orbit round the Sun as members of the solar system, and are not merely passing strangers paying fleeting visits. The reason why this simple fact had not been appreciated before is that there is often an exceedingly long time between successive reappearances of a comet. Halley had been impressed by the spectacle of a comet which was seen in 1682. He made systematic observations of the comet's movement across the sky, and, with a good deal of help from Newton, was able to calculate the shape of its orbit, which he found to be an extremely elongated ellipse.

Once he knew that the comet was moving on an elliptical orbit, he realized that it *had* to come back around again, and, more important, that it must have been seen on previous occasions. He soon spotted in historical records that the same comet – *his* comet – had been seen in 1531 and 1607, and he predicted that the comet would reappear again in 1758. Sadly, he died before he could see his prediction confirmed, but other astronomers developed Halley's work, and they confirmed his prediction. In fact, the comet was first spotted by a farmer with a home-made telescope on Christmas Day 1758. Incidentally, the detailed work on the motions of Halley's Comet, particularly by the French astronomer Clairaut, laid many of the foundations of modern celestial mechanics.

Figure 10.15 shows just how elliptical the orbit of Halley's Comet is. The figure also illustrates one of the main problems Halley had to deal with in his calculations: the shape of the orbit near the Sun is extremely close to that of a parabola, which is an open figure. Halley's Comet belongs to a group which are called *short-period* comets; their orbits are elliptical, and they return relatively frequently. *Long-period* comets, by contrast, move on orbits which are so nearly parabolic that even today, with the best

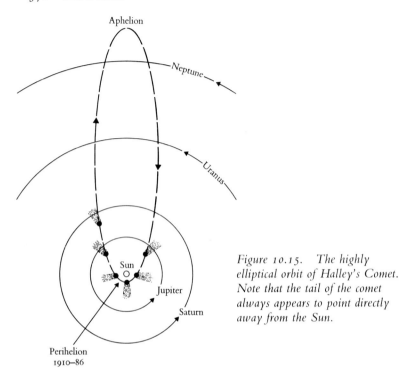

Figure 10.15. The highly elliptical orbit of Halley's Comet. Note that the tail of the comet always appears to point directly away from the Sun.

equipment, it is very difficult to determine their orbital character-istics, and many of them are thought to have periods of thousands or tens of thousands of years. So, if you were lucky enough to see one of these long-period comets, it is a depressing certainty that you will *never* see it again.

Although their orbits *appear* to be parabolic, it is thought the long-period comets do follow elliptical orbits. This means that they swing out to *enormous* distances from the Sun before begin-ning their return journeys. Most of them reach well out into the depths of space beyond Pluto. Some of the most spectacular comets are those whose orbits take them extremely *near* to the Sun. These are called 'sungrazers', appropriately enough, since some of them seem almost to brush through the outer fringes of the Sun's atmosphere.

Comets, then, are rather shy objects, spending most of their time in the remote, chilly parts of the solar system. Just occasionally they come cautiously in towards the Sun and then hurry away again, like a timid wild bird accepting an offering of food from a human. For most of the time, any individual comet is quite invisible from the Earth. This raises an interesting question. How many comets are there Out There? If a comet does not come nearer the Sun than, say, the orbit of Saturn, it would be perpetually invisible. Estimates of the numbers of comets vary enormously, but most are on the high side, ranging up to many millions. J. H. Oort has suggested that there may be a vast 'cloud' of comets surrounding the solar system, well out beyond Pluto. Of these, only a few will ever become visible, when a chance perturbation sends them sunwards. Although a huge number of comets may exist in the cloud, the total mass of material may be quite small, perhaps only a few times the mass of the Earth.

The majority of comets, then, remain for ever invisible to us. In the course of a lifetime, however, an individual should be able to see one or two great examples sweeping past the Earth: Halley's Comet is due to make its next appearance in 1986. It is also possible that, in the course of a lifetime, readers may find themselves being swept right *through* a comet, if the orbits of a comet and the Earth intersect at the right time. Fortunately, this eventuality is not nearly so unpleasant as might be imagined.

Although no two comets ever look quite the same, they usually have two things in common: a small, indistinct *head*, and a great long banner of a *tail*. It is often said that a woman's crowning glory is her hair; the great glory of a comet is certainly in its tail, which may stream across the entire vault of the night sky. (Coincidentally or not, the word *comet* comes from the Greek word for *hairy*; the Greeks thought of comets as hairy stars.) The tail may extend for tens of millions of kilometres, but the head may be quite tiny, perhaps only a few kilometres in diameter.

Comets are far from being fully understood, but Fred Whipple, the doyen of comet workers, has suggested an apt analogy: a comet is a large, dirty snowball in space, spouting a celestial fountain. The snowball is the head. Spectroscopic studies have

Figure 10.16. Comet Cunningham, photographed on 21 December 1940. The tail is rather small but the head is well defined. Note the trails left by the stars in the photograph, indicating the difference in relative motion between the comet and its background. (Hale Observatory photograph)

shown that the materials making up the head are probably water ice, together with frozen gases such as methane and ammonia, and perhaps small quantities of dusty or stony silicate materials. We cannot yet directly sample material from the celestial snow-balls, though space missions are on the drawing board which will be able to, but it is possible that some of the meteorites which strike the Earth contain a clue. The meteorites concerned are the carbonaceous chondrites. When seen on Earth, they do not, of course, contain any ice or frozen gases: there is no way in which these could survive the incandescent heat of their descent through the Earth's atmosphere. Before they encountered the atmosphere, some may have been much richer in volatiles, and may even have been themselves small snowballs. An indirect pointer to this is that some carbonaceous chondrites arrive on Earth in a decomposed state, sometimes only as loose agglomerates of dust, strongly suggesting that they have lost an icy component. One could say that, of an original dirty snowball, only the dirt arrives.

It would be misleading to imply, however, that the head of a comet is a huge solid mass of ice, containing lumps of solid material. The density of the head is very low, so the ice and rocky material is probably present in small lumps, loosely aggregated together in a moving cloud.

If the head of a comet is a little lacking in substance, then its glorious tail is even more so: there is almost nothing to it! This statement is not wholly facetious. Though comet tails occupy huge volumes, their density is not much above that of empty space.

When a comet approaches the Sun, some of its ice and gas gets vaporized by solar radiation and is driven off by solar wind in a long plume down-Sun. (This is why the tail of a comet always points away from the Sun.) Some dusty, solid particles are carried away as well, and these are lit up by the Sun to make the tail visible. In addition to this direct illumination, some of the gas is ionized by the Sun and glows of its own accord, like the gas in a fluorescent light tube. Naturally, this greatly enhances the appearance of the tail.

In detail, two components can be discerned in a comet's tail. One consists of tiny 'ordinary' particles stripped off from the head by radiation pressure, and following the ordinary laws of motion for orbiting bodies. They tend to form part of the tail which curves markedly behind the comet. The other component consists of ionized gas molecules which are caught up in the blast of the solar wind, which generates a magnetic field in the vicinity of the comet and which can accelerate the ionized molecules to velocities of several tens of kilometres per second, while leaving the 'ordinary' particles unaffected.

When a comet recedes from the Sun and the effect of the solar blow-lamp diminishes, the tail fades and the head drifts on drably as a deeply frozen dirty snowball, to remain invisible for long ages. In its passage, however, the comet leaves behind a lot of the dirt blasted off it, and, since many comets come between Sun and Earth, it sometimes happens that the Earth moves through the debris left hanging in space by a long-gone comet. Sometimes, instead of single shooting stars being seen, they appear to arrive in great showers from a single point in space. One of these showers takes place in late April from the constellation Aquarius: it is believed to be related to the passage of Halley's Comet. The most amazing meteor shower of recent times took place in Arizona on 17 November 1966, when over 1,000 shooting stars *a minute* could be seen for nearly an hour, so that the whole sky was lit up. Those lucky enough to witness this display could not find superlatives strong enough to describe what they had seen: it was the experience of a lifetime. Although enormously impressive, this meteor shower was associated with a rather minor comet, one which had been recorded by the Chinese in 1366 but which had been 'lost' for a long time.

So, passing through the tail of a comet is something to be looked forward to rather than dreaded. An encounter with the head, a great dirty snowball, could be unpleasantly different. It's been suggested that the great Tunguska event of 1908 was due to a collision between the Earth and a comet. A large fireball was seen, massive explosions were heard and trees were felled within 30 kilometres radius from 'ground zero' in a remote part of Siberia. Unfortunately, disturbances in Russia – including the Russian Revolution and other trifling events – prevented scientific examination of the area until 1927, nineteen years after the impact, when most of the evidence had been destroyed. The main reason for suggesting that the 'fireball' was caused by the arrival of a comet rather than a solid meteorite is that no well-defined crater was formed: the body seems to have disintegrated well before it hit the ground. Whether it was a small comet or a large 'soft' meteorite is rather an academic point. What is certain is that a large comet impacting on the Earth could prove to be a rather unwelcome visitor.

11. The Origin of the Planets

In this final chapter, we address ourselves to one of the most fundamental problems in science. In doing so, we leave the realms of straightforward observation and deduction and enter an area where physics and philosophy, relativity and religion all meet head on. We shall only tiptoe a short way into this contested ground, however, and will not even approach the main battlefield, where ideas on the origin of the universe come into conflict – the origin of the solar system presents quite enough problems!

A French philosopher once remarked that he did not have time to be brief. Trying to review the origin of the solar system presents a similar problem: there are so many separate facets to take into account, and so many different interpretations. It is almost impossible for a single human mind to grasp all the concepts involved. This is reflected in scientific writings on the subject – an isotopic chemist has a quite different view of the origin of the planets from a theoretical physicist. One wonders sometimes if they are writing about the same event. To make matters worse, there has been a distinct lack of solid evidence to work on in the 300 years during which the problem has been seriously investigated, so scientists have been confined to endless hypothesizing. In this respect, the last few years have seen a drastic change for the better.

A comprehensive review of the origin of the planets would take in everything from meteorites to magnetic fields, and from asteroids to atmospheres. This would require a separate volume to itself, so here only some of the most important aspects of the problem are discussed. These are:

The age of the solar system
The distribution of angular momentum
The variation in composition between the planets
The distribution of mass in the solar system
The origin of the layered structure of the planets

The *age* of the solar system is relatively easy to dispose of; the other topics require more extensive analysis. You will recall from earlier sections that the origin of the solar system is confidently taken to be about 4·6 billion years, since this is the date obtained independently from radiometric dating of samples from the Earth, Moon and meteorites. The *first* events in the formation of the planets took place shortly before 4·6 billion years ago, as we shall see. But now we must address ourselves to the major question: what singular event was it that took place 4·6 billion years ago to give birth to the Earth and the other members of the Sun's family?

The starting point for all thinking on the origin of the solar system is generally taken to be an amorphous cloud of dust and gas at the centre of which the Sun eventually emerged. This shapeless blob in space is called the *primitive solar nebula*. We have to try and explain how this disorganized cloud came to sort itself out into a neat set of planets, asteroids, comets and meteorites. Why did it not just remain an amorphous cloud?

The most difficult part of this problem has taxed scientists for hundreds of years: the distribution of angular momentum in the solar system. Ordinary momentum is relatively straightforward: the momentum of a body is its mass times its velocity. A super-tanker coasting along at one knot has much more momentum than a sailing dinghy spanking along at ten knots; it takes several kilometres to bring a supertanker to a halt. Momentum cannot be destroyed: when a billiard ball hits another it may come to rest itself, but its momentum is transferred to the second ball.

Angular momentum is identical, except that it is concerned with rotation rather than linear movement. A spinning top has angular momentum. It is much more difficult to stop a heavy top from spinning than a light one revolving at the same speed.

There are many different bodies in the solar system, and each one is both rotating on its own axis and revolving round the Sun: each has some angular momentum. The chief difficulty that theories for the origin of the solar system have to overcome is this. The Sun is much the most massive body in the solar system, but it rotates so slowly (once every twenty-seven days) that it has only a small fraction (2 per cent) of the total angular momentum of the solar system. This is counter to intuition: one might expect the Sun, being at the centre of the system, to rotate faster than anything else. Consider an ice skater pirouetting on the tip of one skate. If she extends her arms, she will spin more slowly; if she brings her arms close to her body, she will spin faster – a technique often used for effect by ice dancers.

It was René Descartes who first proposed in 1644 the existence of a primitive solar nebula which would become disc-like through rotation. This idea was elaborated by his compatriot Pierre Laplace, who realized that a cloud of dust and gas, shrinking to form the Sun, would – like the ice skater – have to spin faster and faster in order to conserve angular momentum. Laplace suggested a series of rings were shed, from which the planets and their satellites later formed. Although Laplace's work was widely regarded for many years, it did not fully account for the rotations of the individual planets, or for their angular momenta. In the centuries after Laplace, many different ideas were tossed around. These alternated between ones which involved only *one* body (the primitive solar nebula) and those that required a *second* body to take part. The most popular of these involved a near collision between the proto-Sun and another passing star. Tidal interactions between the two bodies would have resulted in a long filament of nebular material being ripped out of the Sun, from which the planets formed. Although now rejected for rather complex mathematical reasons, these two-body theories also carry the implication that planet-forming events must be rather rare in the history of the universe. Many modern astronomers think that the reverse may be the case.

Until recent years, the most popular model for the origin of the planets was formulated by the German astrophysicist, C. von

Weizsäcker. He advocated a single solar nebula, in which a complex set of turbulent vortices was set up. Planets were supposed to form where vortices interacted. This model accounts for the distribution of the planets fairly well, but it seems a bit too regular and mathematically elegant to be plausible, and does not offer a convincing explanation of how transference of angular momentum from the centre outwards took place.

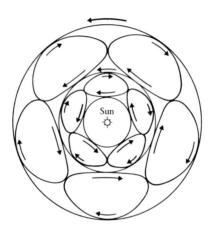

Figure 11.1. Von Weizsäcker's elegantly geometrical model for the origin of the planets.

We still do not really know what happened to turn the disorganized gas and dust of the primitive nebula into the Sun and planets. But one important new factor emerged in 1976. You may recall that 2 tonnes of carbonaceous chondrite meteorite fell near Pueblito de Allende, Mexico, in 1969. This was almost literally a heaven-sent gift to planetologists, as it proved to contain some major clues to the origin of the planets. Since it fell, the meteorite has been subjected to minute chemical investigation. In 1976, it was reported that it contained unusual amounts of isotopes such as magnesium 26.

An element may have several isotopes. Chemically, an element's isotopes are identical, but they differ in the number of neutrons that exist in the nucleus of the atom. Magnesium 26 has twelve protons and fourteen neutrons in its nucleus; other isotopes exist with the same number of protons, but with twelve or thirteen neutrons in the nucleus.

The magnesium 26 in the Allende meteorite is believed to have been formed by the radioactive decay of a quite different isotope, aluminium 26. Now we reach the critical point. Aluminium 26 is quite extinct, as dead as a dodo. There simply is not any more of it in the solar system, because its half-life is so short (720,000 years) that it has all long since decayed away.

So where did the aluminium 26 that gave rise to magnesium 26 come from in the first place? Much the best places to manufacture elements in the universe are so-called *supernovae*, colossal explosions of stars that are known to happen from time to time and which are sometimes big enough to create brilliant objects in the night sky. One of the best known of these 'new stars' is that of 1054, when a star literally blew itself to glory and blazed so brightly that it was visible in daylight, to the consternation of the Chinese and Japanese astronomers who observed it.

It is possible that a spectacular supernova may have provided a 'trigger' for the origin of the solar system. Assuming that a supernova blew up near the primitive solar nebula, it could not only have 'seeded' the nebula with unmistakable isotopes such as magnesium 26, but the shock waves from it may also have churned up the gas and dust of the nebula in such a way as to disturb the original equilibrium and for planetary nuclei to form. Studies of other isotopes have been used to argue that there may have been *two* separate supernova eruptions shortly before the formation of the solar system, and also that the solar system may have formed directly from the debris of a single supernova, of which the Sun is a dim relic.

While there is no agreement on exactly *how* supernovae were involved in the formation of the solar system, there is general agreement that they must have played a part. It also seems certain from the isotopic evidence that the 'triggering' event happened only a short time before the formation of the planets and meteorites as individual entities – perhaps less than a million years. So, in that short time, the solar nebula had to evolve from an amorphous cloud of dust into the solar system, and sort out its angular momentum problems.

According to the American astronomer, A. G. W. Cameron, the proto-Sun accreted over a period of 30,000 years or so, and while material was falling into the proto-Sun, currents of gas were set up which *blew outwards*, conveying some angular momentum with them. His hypothetical proto-Sun was probably considerably larger than the present Sun, and was surrounded by a thin disc of gas and dust from which the planets formed.

Accretion and layering

The mechanism whereby grains of dust came together to form solid planets is almost as much of a mystery as the first steps in the formation of the solar system, but it is at least more open to investigation and experimentation. Much research has been done in recent years on how particles in space may behave. Having been formed in stars and ejected into space, the grains which formed the starting material in building the planets were probably a millimetre or so in size, and composed of iron, stone and ice with a fluffy texture – possibly tiny 'dirty snowballs'. With billions of these drifting around in space beyond the proto-Sun, it was inevitable that they should collide with one another, and their surface texture made it equally inevitable that they would stick together to form larger grains.

As they milled around in space, these enlarged grains bumped into others from time to time. Sometimes, the collisions caused the grains to break up again into smaller fragments; sometimes they accreted to form larger lumps, perhaps a few millimetres in size. Once they had reached this size, the lumps began to feel the effects of gravity, fell towards the mid-plane of the solar system and interacted with one another. Very rapidly, the billions of separate lumps came together to form hundreds of thousands of asteroid-sized bodies, perhaps larger than a kilometre in diameter. These may have been grouped in rotating clusters with random senses of rotation. The clusters, too, attracted one another gravitationally, and so came together and fused, causing a cumulative increase in the mass of the bodies but not in their angular momentum. As the asteroid-sized bodies grew bigger, so their gravitational attraction got more powerful, and acted over greater distances, so that individual bodies rapidly escalated in size, becoming 'proto-planets'.

The process was undoubtedly vastly more complex than this. For one thing, as the masses of the 'proto-planets' got larger, so too, would the velocities of the in-falling bodies, so that in many cases, violent collisions took place, breaking up the proto-planet into smaller fragments once more: perhaps many of the asteroids and meteorites were formed in this way.

Nor, in this whirling confusion of dirty snowballs and proto-planets, were all the same materials coming together to form identical bodies. Once individual bodies had reached dimensions of more than a few kilometres, compositional differences began to assert themselves. The most important controlling feature here was *temperature*. Although the proto-Sun almost certainly did not shine with as much heat as it did later on, we can safely assume that it was much hotter near the centre of the primitive solar system than at the outside; there must have been a well-established thermal gradient from the centre outwards.

It is to this thermal gradient that the gross differences between the planets can be attributed. Had it not existed, the Earth might have ended up just like a small version of Uranus. Because it was much hotter near the Sun than further out, it was simply not possible for volatile materials like ice and gases to accrete on planets near the Sun, such as Mercury. Mercury was therefore built up of iron and silicate materials which accreted at high temperatures. Moving outwards from the Sun, the planets decrease in density and contain more volatile components. At the distance of Uranus from the Sun, temperatures were low enough for hydrogen to accrete.

This pattern of decreasing temperature away from the Sun and increasing volatile content of the planets is not completely consistent. The Earth is slightly more dense than Venus, whereas

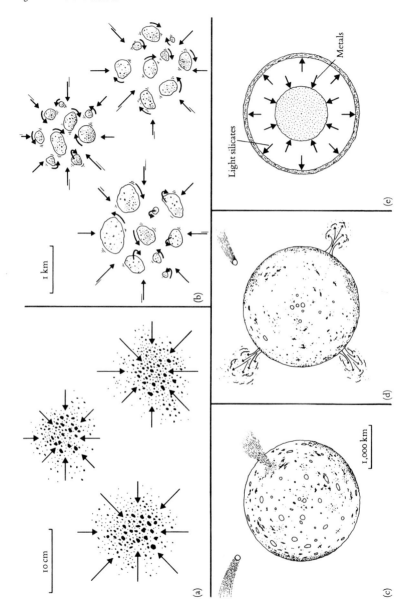

Figure 11.2. A cartoon summarizing the origin of the planets. In (a), tiny dust-like grains are coming together to form small clumps, which are steadily enlarged. In (b), much larger clumps of primitive material have formed, each rotating independently. These come together gravitationally and the clumps fuse, with no net change in the angular momentum distribution. Some of the larger clumps may be broken up by collision with others. In (c), a planetary sized body has formed by accretion of many small fragments. At this point, its composition is controlled by its distance from the Sun: very volatile material would not be included in the accreting grains. Further asteroid-sized bodies are still arriving, and the planet is heating up because of the kinetic energy of all the impacts. In (d), this heating has caused wholesale melting of the planet, and large-scale degassing takes place as volatile gases escape from the centre. Some of these may remain to form an atmosphere; the rest may be lost to space.
In (e), differentiation into core, mantle and crust has commenced, with dense metals sinking towards the core. This process may itself cause further heating.

it should be less dense. This may be because Venus has more oxygen in its mantle. Mars has about the right density for its position, but seems to be poor in volatiles – it should have more than the Earth. The most important components of a planet's budget of volatiles, of course, are to be found in its atmosphere. Both Venus and Earth are well endowed with volatiles, but the present constitution of their atmospheres is largely secondary. Examination of the relative abundances of the so-called 'noble' gases (neon, argon, krypton and xenon), however, reveals that these are present in roughly the same ratios as in carbonaceous chondrite meteorites.

Mars exhibits a similar meteoritic noble gas pattern, but its atmosphere is much less dense and it is clearly depleted in volatiles relative to the Earth by a factor of more than ten. This is the opposite of what might be expected at first, since Mars must have accreted in a cooler part of the solar system than the Earth. The solution lies in its size: accretion is controlled by mass as well as temperature. Thus Mars – and our own Moon – were so small that they were unable to retain volatile elements even at low temperatures. Jupiter and Saturn were, by contrast, so massive that they were able to accrete and retain enormous volumes of hydrogen, the most volatile element of all. In this respect, the two giants can be regarded as 'mini-suns', because the Sun also consists mostly of hydrogen. Furthermore, Jupiter and Saturn are so massive that they would have the same composition no matter where they formed in the solar system; their masses are so great as to completely outweigh the effects of temperature on accretion. Their satellites, however, are small enough to be temperature controlled, so Io and Europa, the smaller Galilean satellites, contain far less volatiles than Ganymede and Callisto.

So far this review of the origin of the solar system has touched on – albeit briefly – three of the five aspects mentioned at the beginning of this section: the age, distribution of angular momentum and variation in composition of planets in the solar system. We have yet to explore the distribution of mass, and the origin of the layered structures of planets. The first of these is much the most difficult, and has been an unsolved problem for decades. Why does the distribution of the planets fit in with Bode's Law, even given the inaccuracies of the outermost members? Is it just coincidence? And why should there be one enormous planet (Jupiter) and several smaller ones, rather than a family of similar-sized bodies? There is no answer to these questions at the moment. Research is being done to establish whether or not our planetary system is in any way special, but recent results indicate that it is only one of a number of different possible types of system.

There is more, much more, to be said about the layering of the planets. The starting point in this discussion is the fact that our own Earth is thought to have a *bulk* composition similar to that of chondritic meteorites, and is thought to have accreted from material of the primitive solar nebula with a composition close to that of carbonaceous chondrite meteorites. But it does not take great powers of observation to deduce that the rocks that we now see at the surface are in no way like chondritic meteorites. Geophysical studies show that the Earth has a metallic core, a mantle of dense silicate minerals and a crust of lower-density silicate minerals. It is only by adding together these three components and averaging them that we arrive at the so-called 'chondritic Earth'.

But how and when did this layered structure develop? And did the Earth experience the same processes as other planets? It used to be thought that all the planets accreted into cold, homogeneous lumps, and that layering developed at a later stage when the planets were warmed up by decay of radioactive elements. Although some layering certainly originated in this way, it is now thought that the planets may have accreted inhomogeneously, in a series of layers, while relatively *hot*, since it would otherwise not be possible to account for the variations in composition *between* planets, which fits in so well with the concept of progressively lower temperatures from the Sun.

Most debate now centres not on whether the planets accreted in a hot or cold state, but how the temperature of the solar nebula varied with time. If the nebula cooled slowly while the planets were accreting, one would expect to find quite a different kind of layering being produced from that which would result if it cooled rapidly.

In both cases, the first materials to condense would be the least volatile 'refractory' compounds, such as aluminium oxide and calcium oxide, to be followed by slightly more volatile iron and nickel, with material such as water and methane condensing much later. But if the temperature of the nebula decreased *slowly*, there would be time for chemical reactions to take place between some of the accreting components – for example, metallic iron could be corroded by hydrogen sulphide gas to form the mineral troilite, iron sulphide. If the cooling were *rapid*, there would be no time for such reactions, and the planet would end up with an 'onion-skin' structure as successive layers of compounds, stable at progressively lower temperatures, rapidly built up.

Although there is a fair amount of evidence in favour of the slow-cooling hypothesis, it is not particularly important for our purposes to argue the merits of each. It is much more useful to consider how each of the members of the solar system relates to the model for the origin of the solar system that we have been developing. Some of the differences between the individual planets result from original differences in what they accreted, some from post-accretion differentiation processes. On the Earth, at least, these processes are still continuing.

Post-accretion processes

Let us start with the simplest, smallest bodies. It should be apparent by now that comets – those dirty snowballs – are probably extremely primitive bodies, lumps of material that came together right at the beginning of the solar system and have been wandering around in space ever since, having been lucky enough to escape banging into any other body. Many meteorites, particularly the carbonaceous chondrites, are probably equally primitive. Others, particularly irons and stony irons, were derived from the larger, asteroid-sized bodies, within which post-accretion internal differentiation had taken place. These asteroids belong to a group of proto-planets which were probably too small to have accreted different layers initially, but did get hot enough to melt, partly through the process of accretion itself, and partly through radioactive heating. This enabled the denser, metallic material to separate into cores, while the less dense silicate materials accumulated around the cores. Many of the larger asteroids we now see probably have exactly this simple structure; others were shattered by collisions to produce the range of meteorite types that end up on Earth – the cores contributed the 'irons' and the outer layers the stony meteorites.

Figure 11.3.   The origin of meteorites and the nature of asteroids.
(a)   Meteorites which arrive on Earth may result from collision between asteroids in the asteroid belt. The collisions may cause particles to be ejected on highly elliptical orbits, enhanced by perturbations by Jupiter.
(b)   A differentiated asteroid after a collision may have a large chunk of itself missing, but much material ejected in the collision will fall back, forming a blanket of fragmented material covering the entire surface.
(c)   Successive collisions in the crowded asteroid belt may reduce the whole asteroid into a jumbled heap of loose fragments. It is unlikely, however, that significant loss of mass will take place, since most fragments will fall back gravitationally after impact. Many asteroids are therefore probably orbiting heaps of broken boulders.

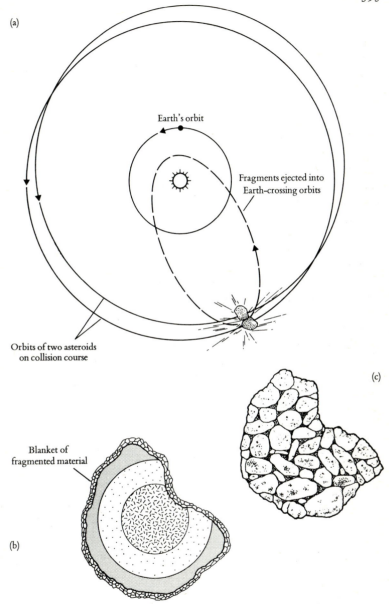

(a)

Earth's orbit

Fragments ejected into
Earth-crossing orbits

Orbits of two asteroids
on collision course

(c)

Blanket of
fragmented material

(b)

Turning now to the planets and their satellites, consideration of their origin helps us to unify some of the many apparent differences between them. Mercury, closest to the Sun, accreted in the hottest environment of all; hence its high density and complete lack of volatiles. It almost certainly experienced large-scale melting after accretion, refining the primitive core into the present massive metallic mass, and a thin crust probably formed at the surface. What happened after that we have yet to learn. Some volcanic activity may have taken place, but Mercury's surface seems to record the impacts of the last fusillade of bombardment in the accretion process, and the planet has probably been quite inactive since about 4 billion years ago. The core may still be hot, perhaps even liquid – it is not possible to say more.

Venus was far enough away from the Sun to retain far more volatiles than Mercury, as its massive atmosphere testifies. Since it formed near the Earth, it should have an internal structure and composition broadly similar to the Earth's. In particular, it should have accreted a similar quantity of radioactive elements, and thus should have experienced a similar history of melting and internal differentiation. The key questions on Venus concern the present state of its interior. Does it possess a mobile mantle like the Earth's? If not, why not? If it does, does it exhibit the same kind of large-scale plate tectonic processes at its surface? Are there drifting continents on Venus? As we saw in Chapter 5, the meagre evidence available so far suggests that it has an old, static, cratered surface; but until we can lift the veil of cloud around the planet and study the surface more closely, it will be impossible to be certain. Venus remains the most enigmatic and interesting planet in the solar system. Many spacecraft will be visiting it in the next decade. There are certain to be some great discoveries.

The Earth accreted at just the right distance from the Sun to provide a congenial environment for us – a little closer, and liquid water could not exist at the surface. It is important to emphasize, however, that the oxygen-rich atmosphere that we breathe in no way resembles the primordial atmosphere that the Earth collected from the primitive solar nebula. The most volatile components of this – such as hydrogen – could not be retained at all, and other volatiles were probably driven off by the Earth's internal heat. Our present atmosphere accumulated slowly as a result of volcanic degassing of the mantle, which sprayed out quantities of steam and carbon dioxide. The all-important oxygen did not become abundant until plant life had evolved which was capable of breaking down carbon dioxide by photosynthesis.

The Earth's internal structure is naturally known in far more detail than any other planet's. The physical properties of the core, mantle and crust have been minutely documented by years of seismological investigations, but two key issues remain hotly debated. How did the layered structure develop, and what is the core made of?

The core and mantle could have been formed either as primary or secondary features. The 'primary' school of thought argues that the iron and nickel of the core accreted first from the primitive solar nebula, and formed the kernel of the proto-Earth. Later, temperatures fell, and silicate materials accreted around the already formed metallic core to make the mantle. The 'secondary' school argues that the material from which the proto-Earth accreted was homogenous, and that the intense heating resulting from the accretion process caused the metallic component to segregate out and settle towards the centre, just as iron settles to the bottom of a blast furnace, leaving the slag on top. It is not easy to choose between these two models, but the difference is rather academic – it is generally accepted that even if the core is of secondary origin, it must have formed extremely rapidly after the accretion process had ceased, perhaps only a few million years.

We have taken it for granted in previous discussions that the core, however formed, is a mixture of iron and nickel. While this must be broadly correct, it turns out that, in detail, the density of the core is definitely too low for it to be purely iron and nickel; there *must* be between ten and twenty per cent of a lower density component present. The nature of this component has been acrimoniously debated. Candidates at present being fielded range from oxygen through sulphur to potassium! The involved geochemical arguments need not detain us here, but it will be necessary to resolve the question of core composition before we can hope to understand fully the way the Earth and other planets work. For example, if there was potassium rather than oxygen in the core, the heat produced by its radioactive isotope would be fully capable of supplying *all* the energy required to drive plate tectonic processes. If oxygen were present in the core, a quite different heat source has to be postulated: the small traces of radioactive elements present in the mantle.

The Earth's interior is, and has always been, mobile, thanks to the heat liberated by its content of radioactive elements. This leads to continual changes in its surface, as the continents shift around and the continental crust steadily thickens. This mobility is in marked contrast to any other planet, and is especially important in relation to Venus. An explanation for this contrast has been sought in the existence of the Moon. Early in the Earth's history, the Moon was much nearer than it is now. Tidal forces between it and the Earth must have been considerable, and it is possible that the heating caused within the Earth by the dissipation of this tidal energy may have been sufficient to ensure that the Earth's interior remained mobile. Lacking a moon of its own, Venus would be denied this *extra* heating, and thus its interior may never have been so thoroughly stirred up.

Although the Moon probably formed near the Earth, it is quite different. It is not massive enough to retain any volatiles, and those that may have been accreted initially were driven off when it experienced post-accretion melting and differentiation. The Moon did acquire some radioactive elements, but only sufficient to keep volcanism going until about 3 billion years ago. Although the centre is still hot, externally it is quite dead.

Mars formed close enough to the Sun to accrete enough refractory elements to qualify as one of the rocky 'terrestrial' planets, but was only able to form a low-density core consisting mostly of iron sulphide. Mars must also have accreted substantial quantities of radioactive elements – witness the huge volcanoes – but its internal heat has probably passed its peak. Its crust is now so thick that volcanic activity can only continue at a very feeble level, if at all, though there may still be a mobile mantle. Its satellites, Phobos and Deimos, appear to be merely chunks of carbonaceous chondrite material. While they resemble 'captured' asteroids in many respects, it is possible that they have always been satellites of Mars and are merely left-over chunks of the primitive material of the solar nebula.

Further out from the Sun, the two giants, Jupiter and Saturn, were so massive that they could accrete and retain everything that came their way. (Why they should have had the potential to grow so large is another, unanswerable question.) As we have seen, the satellites of the two giant planets are in a sense more planet-like than their parents, so that Jupiter's satellites have accreted volatiles in proportion to their masses. Saturn's largest satellite, Titan, is so distant from the Sun, and so massive, that it was even able to accrete and retain its own atmosphere. Of most interest to planetary scientists is the amount of radioactive elements these distant icy satellites were able to accrete. If significant quantities are present, then watery kinds of 'volcanic' processes may be operating on them, as the Voyager pictures of Ganymede have already indicated.

Further out still, our knowledge becomes insufficient to say much about the origins of either the planets or their satellites. Both Uranus and Neptune probably contain large amounts of hydrogen, indicating extremely low accretion temperatures, but they also contain ice, frozen gases and some rock material. Pluto is so small that it cannot have been able to retain hydrogen; so it probably consists only of frozen gases, such as methane, surrounding a tiny rocky core.

★

Here we must end this brief review of the solar system. This book has attempted to cover a huge scope: the nine planets and their attendant satellites; comets, meteorites and asteroids; the work of astronomers and planetologists over several centuries; the Apollo landings and many spacecraft missions. The treatment has necessarily been patchy and incomplete. It is to be hoped, however, that readers will have gleaned something useful from it, even if it is only an awareness of how much remains to be discovered. The last decade has been an extraordinarily enriching one for any mortal with a vestige of curiosity about the Earth and its neighbours in space: it has been a glorious decade for science. If the missions that were *en route* through space while this book was being written prove to be only half as successful as their predecessors, the next decade promises to be just as golden. Watch this space!

Suggestions for Further Reading

1. Historical Background

Arthur Koestler, *The Sleepwalkers*, Hutchinson, London, 1959. Discusses the evolution of ideas on the solar system in a stimulating, lively style.

W. Ley, *Watchers of the Skies*, Sidgwick & Jackson, London, 1964. A fascinating account of the history of astronomy.

E. Lessing, *Discoverers of Space*, Burns & Oates, London; Herder & Herder, Vienna and Freiburg, 1969. A well-illustrated 'coffee-table' book dealing with some of the great astronomers.

2. The Solar System

The Solar System, a Scientific American book, W. H. Freeman & Co., San Francisco, 1975. Probably the best brief summary of modern views on the solar system and its origin.

G. H. A. Cole, *The Structure of Planets*, Wykeham Publications, London, 1978. A specialist book for the mathematically minded, concerned with the determination of the internal structures of planets.

3. The Moon

Bevan M. French, *The Moon Book*, Penguin Books, Harmondsworth, 1977. A very readable account of the scientific aspects of the Apollo project.

S. R. Taylor, *Lunar Science: a Post-Apollo View*, Pergamon Press, Inc., New York, 1975. An authoritative technical account of the geological and geochemical aspects of the Apollo project.

J. E. Guest and R. Greeley, *Geology on the Moon*, Wykeham Publications, London, 1978. A review of lunar geology aimed at undergraduate readers.

T. A. Mutch, *The Geology of the Moon: a Stratigraphic View*, Princeton University Press, 1970. Although written before the Apollo landings, this book presents a clear, well-illustrated summary of the methods used in investigating the Moon. Lavishly illustrated with Lunar Orbiter pictures.

4. Mars

P. Lowell, *Mars as the Abode of Life*, Macmillan & Co., New York, 1908. Not easily obtainable, but well worth reading. Contains full accounts of Lowell's ideas about the canals of Mars and their builders.

W. G. Hoyt, *Lowell and Mars*, University of Arizona Press, 1976. An excellent biography of a remarkable man.

Mars as Viewed by Mariner 9, NASA Special Publication No. 329, 1974. A photographic summary of the results of the mission, with some explanatory text. Gives a good general guide to the geography of Mars.

By the Viking Lander Imaging Team, *The Martian Landscape*, NASA Special Publication No. 425, 1978. A presentation of the best Viking lander pictures, backed up with technical discussion of the features seen in the pictures. Contains an excellent account of the camera design, and some very revealing insights into the background to the whole mission.

Scientific Results of the Viking Project, vol. 82 of the *Journal of Geophysical Research*, 1977 (published by the American Geophysical Union). Contains more than fifty technical papers dealing with all aspects of the Viking missions, presenting *all* the preliminary results.

5. Planetary Satellites

J. A. Burns (ed.), *Planetary Satellites*, University of Arizona Press, 1977. A collection of papers representing the proceedings of a conference, this book contains a unique compilation of physical data on the satellites and is the only volume of its kind.

6. Planetary Geology

E. A. King, *Space Geology*, John Wiley & Sons, Chichester, 1976. Rather a misleading title for a book which is mostly concerned with meteorites and the Moon, but none the less a good introduction to those interested in impacts and their effect on rocks.

N. M. Short, *Planetary Geology*, Prentice-Hall, Englewood Cliffs, N.J., 1975. Also strongly orientated towards the Moon, this book does provide a rather broader based geological view. Becoming rapidly out-dated by events.

The books mentioned above are only a tiny fraction of the vast literature on the planets, but appear to the author to present a good introduction to the subject. It is inevitable that books on such a rapidly evolving subject soon become out-dated. Readers who wish to keep abreast of modern thinking are strongly recommended to consult such periodicals as *Scientific American* and *Sky and Telescope*, both of which frequently have excellent review articles concerning recent discoveries, written in a lucid, non-technical style.

Index

More About Penguins
And Pelicans

For further information about books available from Penguins please write to Dept EP, Penguin Books Ltd, Harmondsworth, Middlesex UB7 0DA.

In the U.S.A.: For a complete list of books available from Penguins in the United States write to Dept CS, Penguin Books, 625 Madison Avenue, New York, New York 10022.

In Canada: For a complete list of books available from Penguins in Canada write to Penguin Books Canada Ltd, 2801 John Street, Markham, Ontario L3R 1B4.

In Australia: For a complete list of books available from Penguins in Australia write to the Marketing Department, Penguin Books Australia Ltd, PO Box 257, Ringwood, Victoria 3134.

Some Other Pelican Books

EARTH'S AURA
A Layman's Guide to the Atmosphere
Louise B. Young

'It must be one of the best popular science books which deal with the atmosphere.' – *Nature*

CONTINENTAL DRIFT
A Study of the Earth's Moving Surface
D. H. and M. P. Tarling

'. . . a model exposition: succinct, but not dodging difficulties; eschewing jargon; most helpfully illustrated.' – *Daily Telegraph*

DISASTERS

John Whittow

During the last four or five years rarely a month has gone by without a major disaster being reported from somewhere in the world.

What is going on? Is it possible to predict disasters or to alleviate their effect? John Whittow not only gives a detailed description of the disasters which have caused such a widespread havoc, but also explains as far as possible why they occur and whether man is partly to blame.

TIME, SPACE AND THINGS

B. K. Ridley

'This book is an attempt to survey, in simple terms, what physics has to say about the fundamental structure of the universe. It aims to extract, from a whole range of specialized activities, the basic essential concepts and to present them in plain, non-mathematical language. There are some splendidly bizarre ideas in physics, and it seems a pity to keep them locked up in narrow boxes, available only to a small esoteric crowd of key-holders.'

Another Pelican Original by Peter Francis:

VOLCANOES

What is a volcano? And where, how and why does volcanism exist? Are eruptions predictable, and how dangerous, or even beneficial, are they? How do they affect the environment and influence climatic conditions?

This is a clear and detailed book which fully answers these questions and describes the volcanic phenomenon in all its aspects. With eye-witness accounts, ranging from Vesuvius in A.D. 79 (the younger Pliny) to Krakatoa in 1883, and other well-documented terrestrial and sub-marine instances, Dr Francis has produced an up-to-date and absorbing study, often surprising in its conclusions and always thought-provoking.